The Creation of Strange Non-Chaotic Attractors in Non-Smooth Saddle-Node Bifurcations

MEMOIRS
of the
American Mathematical Society

Number 945

The Creation of Strange Non-Chaotic Attractors in Non-Smooth Saddle-Node Bifurcations

Tobias H. Jäger

September 2009 • Volume 201 • Number 945 (fourth of 5 numbers) • ISSN 0065-9266

American Mathematical Society
Providence, Rhode Island

2000 *Mathematics Subject Classification.*
Primary 37D45; Secondary 37C60, 37H20.

Library of Congress Cataloging-in-Publication Data

Jäger, Tobias H., 1976–
The creation of strange non-chaotic attractors in non-smooth saddle-node bifurcations / Tobias H. Jäger.
p. cm. — (Memoirs of the American Mathematical Society, ISSN 0065-9266 ; no. 945)
"Volume 201, number 945 (fourth of 5 numbers)."
Includes bibliographical references and index.
ISBN 978-0-8218-4427-4 (alk. paper)
1. Attractors (Mathematics) 2. Bifurcation theory. 3. Differentiable dynamical systems. I. Title.

QA614.813.J34 2009
515′.39—dc22 2009023713

Memoirs of the American Mathematical Society

This journal is devoted entirely to research in pure and applied mathematics.

Subscription information. The 2009 subscription begins with volume 197 and consists of six mailings, each containing one or more numbers. Subscription prices for 2009 are US$709 list, US$567 institutional member. A late charge of 10% of the subscription price will be imposed on orders received from nonmembers after January 1 of the subscription year. Subscribers outside the United States and India must pay a postage surcharge of US$65; subscribers in India must pay a postage surcharge of US$95. Expedited delivery to destinations in North America US$57; elsewhere US$160. Each number may be ordered separately; *please specify number* when ordering an individual number. For prices and titles of recently released numbers, see the New Publications sections of the *Notices of the American Mathematical Society.*

Back number information. For back issues see the *AMS Catalog of Publications.*

Subscriptions and orders should be addressed to the American Mathematical Society, P. O. Box 845904, Boston, MA 02284-5904 USA. *All orders must be accompanied by payment.* Other correspondence should be addressed to 201 Charles Street, Providence, RI 02904-2294 USA.

Memoirs of the American Mathematical Society (ISSN 0065-9266) is published bimonthly (each volume consisting usually of more than one number) by the American Mathematical Society at 201 Charles Street, Providence, RI 02904-2294 USA. Periodicals postage paid at Providence, RI. Postmaster: Send address changes to Memoirs, American Mathematical Society, 201 Charles Street, Providence, RI 02904-2294 USA.

This publication is indexed in *Science Citation Index*®, *SciSearch*®, *Research Alert*®, *CompuMath Citation Index*®, *Current Contents*®/*Physical, Chemical & Earth Sciences.*
Printed in the United States of America.

∞ The paper used in this book is acid-free and falls within the guidelines established to ensure permanence and durability.
Visit the AMS home page at `http://www.ams.org/`

10 9 8 7 6 5 4 3 2 1 14 13 12 11 10 09

Contents

Abstract

We propose a general mechanism by which strange non-chaotic attractors (SNA) are created during the collision of invariant curves in quasiperiodically forced systems. This mechanism, and its implementation in different models, is first discussed on an heuristic level and by means of simulations. In the considered examples, a stable and an unstable invariant circle undergo a saddle-node bifurcation, but instead of a neutral invariant curve there exists a strange non-chaotic attractor-repeller pair at the bifurcation point. This process is accompanied by a very characteristic behaviour of the invariant curves prior to their collision, which we call *'exponential evolution of peaks'*.

This observation is then used to give a rigorous description of non-smooth saddle-node bifurcations and to prove the existence of SNA in certain parameter families of quasiperiodically forced interval maps. The non-smoothness of the bifurcations and the occurrence of SNA is established via the existence of *'sink-source-orbits'*, meaning orbits with positive Lyapunov exponent both forwards and backwards in time.

The important fact is that the presented approach allows for a certain amount of flexibility, which makes it possible to treat different models at the same time - even if the results presented here are still subject to a number of technical constraints. This is unlike previous proofs for the existence of SNA, which are all restricted to very specific classes and depend on very particular properties of the considered systems. In order to demonstrate this flexibility, we also discuss the application of the results to the Harper map, an example which is well-known from the study of discrete Schrödinger operators with quasiperiodic potentials. Further, we prove the existence of strange non-chaotic attractors with a certain inherent symmetry, as they occur in non-smooth pitchfork bifurcations.

Received by the editor 7th Febuary 2006.

2000 *Mathematics Subject Classification.* Primary 37D45, Secondary 37C60, 37H20.

Key words and phrases. Strange non-chaotic attractors, quasiperiodically forced systems, non-autonomous bifurcations.

This work was supported by the German Research Foundation (DFG), grant Ke 514/6-1.

CHAPTER 1

Introduction

In the early 1980's, Herman [**1**] and Grebogi et al. [**2**] independently discovered the existence of strange non-chaotic attractors (SNA's) in quasiperiodically forced (qpf) systems. These objects combine a complicated geometry[1] with non-chaotic dynamics, a combination which is rather unusual and has only been observed in a few very particular cases before (the most prominent example is the Feigenbaum map, see [**3**] for a discussion and further references). In quasiperiodically forced systems, however, they seem to occur quite frequently and even over whole intervals in parameter space [**2, 4, 5**]. As a novel phenomenon this evoked considerable interest in theoretical physics, and in the sequel a large number of numerical studies explored the surprisingly rich dynamics of these relatively simple maps. In particular, the widespread existence of SNA's was confirmed both numerically (see [**6**]–[**19**], just to give a selection) and even experimentally [**21, 22, 23**]. Further, it turned out that SNA play an important role in the bifurcations of invariant circles [**5, 14, 18, 20**].

The studied systems were either discrete time maps, such as the qpf logistic map [**10, 13, 18**] and the qpf Arnold circle map [**5, 9, 12, 14**], or skew product flows which are forced at two or more incommensurate frequencies. Especially the latter underline the significance of qpf systems for understanding real-world phenomena, as most of them were derived from models for different physical systems (e.g. quasiperiodically driven damped pendula and Josephson junctions [**6, 7, 8**] or Duffing oscillators [**22**]. Their Poincaré maps again give rise to discrete-time qpf systems, on which the present article will focus.

However, despite all efforts there are still only very few mathematically rigorous results about the subject, with the only exception of qpf Schrödinger cocycles (see below). There are results concerning the regularity of invariant curves ([**24**], see also [**25**]), and there has been some progress in carrying over basic results from one-dimensional dynamics [**26, 27, 28**]. But so far, the two original examples in [**1**] and [**2**] remain the only ones for which the existence of SNA's has been proved rigorously. In both cases, the arguments used were highly specific for the respective class of maps and did not allow for much further generalisation, nor did they give very much insight into the geometrical and structural properties of the attractors.

The systems Herman studied in [**1**] were matrix cocycles, with quasiperiodic Schrödinger cocycles as a special case. The linear structure of these systems and their intimate relation to Schrödinger operators with quasiperiodic potential made it possible to use a fruitful blend of techniques from operator theory, dynamical

[1]This means in particular that they are not a piecewise differentiable (or even continuous) sub-manifold of the phase space.

systems and complex analysis, such that by now the mathematical theory is well-developed and deep results have been obtained (see [**29**] and [**30**] for recent advances and further reference). However, as soon as the particular class of matrix cocycles is left, it seems hard to recover most of these arguments. One of the rare exceptions is the work of Bjerklöv in [**31**] (taken from [**32**]) and [**33**], which is based on a purely dynamical approach and should generalise to other types of systems, such as the ones considered here. (In fact, although implemented in a different way the underlying idea in [**33**] is very similar to the one presented here, such that despite their independence the two articles are closely related.)

On the other hand, for the so-called '*pinched skew products*' introduced in [**2**], establishing the existence of SNA is surprisingly simple and straightforward (see [**4**] for a rigorous treatment and also [**34**] and [**35**]). But one has to say that these maps were introduced especially for this purpose and are rather artificial in some aspects. For example, it is crucial for the argument that there exists at least one fibre which is mapped to a single point. But this means that the maps are not invertible and can therefore not be the Poincaré maps of any flow.

The main goal of this article is to prove the existence of SNA in certain parameter families of qpf systems where this has not been possible previously. Thereby, we will concentrate on a particular type of SNA, namely 'strip-like' ones, which occur in saddle-node and pitchfork bifurcations of invariant circles (see Figure 1.1, for a more precise formulation consider the definition of invariant strips in [**26**] and [**28**]). In such a saddle-node bifurcation, a stable and an unstable invariant circle approach

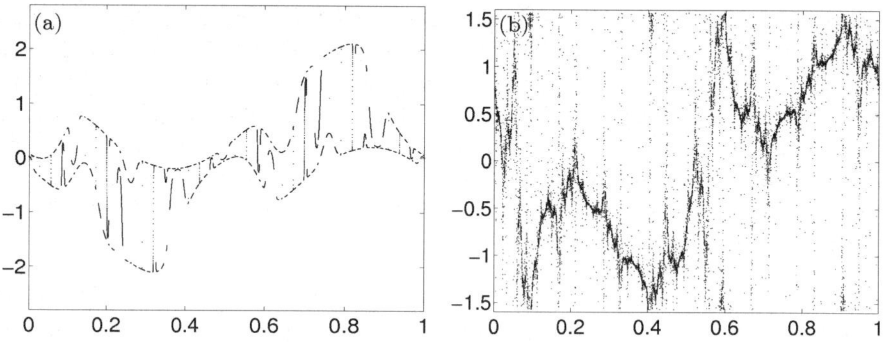

FIGURE 1.1. Two different types of strange non-chaotic attractors: The left picture shows a 'strip-like' SNA in the system $(\theta, x) \mapsto (\theta + \omega, \tanh(5x) + 1.2015 \cdot \sin(2\pi\theta))$. The topological closure of this object is bounded above and below by semi-continuous invariant graphs (compare (1.4)). This is the type of SNA's that will be studied in the present work. The right picture shows a different type that occurs for example in the critical Harper map (Equation (1.6) with $\lambda = 2$ and $E = 0$; more details can be found in [**36**]), where no such boundaries exist. In both cases ω is the golden mean.

each other, until they finally collide and then vanish. However, there are two different possibilities. In the first case, which is similar to the one-dimensional one, the two circles merge together uniformly to form one single and neutral invariant circle at the bifurcation point. But it may also happen that the two circles approach each other only on a dense, but (Lebesgue) measure zero set of points. In this case, instead of a single invariant circle, a strange non-chaotic attractor-repeller-pair is created at the bifurcation point. Attractor and repeller are interwoven in such a way, that they have the same topological closure. This particular route for the creation of SNA's has been observed quite frequently ([**12, 14, 15, 19**], see also [**10**]) and was named *'non-smooth saddle-node bifurcation'* or *'creation of SNA via torus collision'*. The only rigorous description of this process so far was given by Herman in [**1**]. In a similar way, the simultaneous collision of two stable and one unstable invariant circle may lead to the creation of two SNA's embracing one strange non-chaotic repeller [**5, 16**].

Acknowledgments. The results presented here were part of my thesis, and I would like to thank Gerhard Keller for his invaluable advice and support during all the years of my PhD-studies. I am also greatful to an anonymous referee, whose thoughtful remarks greatly improved the manuscript. This work was supported by the German Research Foundation (DFG), grant Ke 514/6-1.

1.1. Overview

As mentioned above, the main objective of this article is to provide new examples of SNA, by describing a general mechanism which is responsible for the creation of SNA in non-smooth saddle-node bifurcations. While this mechanism might not be the only one which exists, it seems to be common in a variety of different models, including well-known examples like the Harper map or the qpf Arnold circle map. The evidence we present will be two-fold: In the remainder of this introduction we will explain the basic idea, and discuss on an heuristic level and by means of numerical simulations how it is implemented in the two examples just mentioned and a third parameter family, which we call arctan-family. An analogous phenomenom is also observed in so-called Pinched skew products, first introduced in [**2**], even if no bifurcation takes place in these systems.

The heuristic arguments given in the introduction are then backed up by Theorem 2.7, which provides a rigorous criterium for the non-smoothness of saddle-node bifurcations in qpf interval maps. This leads to new examples of strange non-chaotic attractors, and the result is flexible enough to apply to different parameter families at the same time, provided they have similar qualitative features and share a certain scaling behaviour. Nevertheless, it must be said that there is still an apparent gap between what can be expected from the numerical observations and what can be derived from Theorem 2.7 . For instance, the latter does not apply to the forced version of the Arnold circle map, and for the application to the arctan-family and the Harper map we have to make some quite specific assumptions on the forcing function and the potential, respectively. (Namely that these have a unique maximum and decay linearly in a neighbourhood of it). However, our main concern here is just to show that the general approach we present does lead to rigorous results at all, even if these are still far from being optimal. The present work should therefore be seen rather as a first step in this direction, which will hopefully inspire further research, and not as an ultimate solution.

The article is organised as follows: After we have given some basic definitions, we will introduce our main examples in Section 1.3 . As mentioned, these are the arctan-family with additive forcing, the Harper map, the qpf Arnold circle map and

Pinched skew products. The simulations we present mostly show the evolution of stable invariant curves as the system parameters are varied. The crucial observation is the fact that the behaviour of these curves prior to the bifurcation follows a very characteristic pattern, which we call *'exponential evolution of peaks'*. In all the first three examples the qualitative features of this process are similar, and even in Pinched skew products, where no saddle-node bifurcation occurs, an analogue behaviour can be observed. Finally, a slight modification of the arctan-family is used to illustrate that the phenomenom is also present in non-smooth pitchfork bifurcations.

On an heuristic level it is not difficult to give an explanation for this behaviour, and this will be done in Section 1.4 . The simple geometric intuition obtained there will then determine the strategy for the rigorous proof of the non-smoothness of the bifurcations in the later sections. More precisely, the heuristics indicate why the existence of SNA should be linked to the appearance of sink-source-orbits in these situations, and this will be one of the main ingredients of the proof.

Section 2 then contains the statement of our main results and discusses their application to the examples from the introduction (or why such an application is not possible, in the case of the qpf Arnold circle map). Before we can turn to the existence of SNA and the non-smoothness of bifurcations, we need to state two preliminary results. The first, Theorem 2.1, provides a general framework in which saddle-node bifurcations in qpf interval maps take place (smooth or non-smooth). The second, Theorem 2.4, states that the existence of sink-source-orbits[2] implies the existence of SNA's (although the converse is not true). After these statements and some related concepts have been introduced in Sections 2.1 and 2.2, we can turn to the main result, namely Theorem 2.7, which provides a criterium for the existence SNA's created in non-smooth saddle-node bifurcations. The counterpart for non-smooth pitchfork bifurcations is Theorem 2.10, which gives a criterium for the existence of symmetric SNA's. More precisely, under the assertions of this theorem there exists a triple consisting of two SNA, symmetric to each other, which embrace a self-symmetric strange non-chaotic repeller. These objects are presumably created by the simultaneous collision of two stable and one unstable invariant curve. However, as the considered parameter families lack a certain monotonicity property which is present in the situation of Theorem 2.7, we cannot describe the bifurcation pattern in a rigorous way as for the saddle-node bifurcations, such that the existence of SNA is the only conclusion we draw in the symmetric setting. The application of these results to the arctan-family and the Harper map is then discussed in detail in Section 2.4, which resumes the structure of Section 1.3 where these examples are introduced. As we have mentioned before, the statement of Theorem 2.7 is too restricted to apply to the qpf Arnold circle map. However, in Section 2.4.3 we discuss some possible modifications, which might allow to treat this example in a similar way, at least for particular forcing functions.

Section 3 provides the proofs for the more elementary results (namely Theorems 2.1 and 2.4). All the remaining sections are then dedicated to the proof of Theorems 2.7 and 2.10, starting with an outline of the construction in Section 4.

---

[2]Orbits with positive Lyapunov exponent both forwards and backwards in time, see Definition 2.3 .

1.2. Basic definitions and notations

A *quasiperiodically forced (qpf) system* is a continuous map of the form

$$T : \mathbb{T}^1 \times X \to \mathbb{T}^1 \times X \quad , \quad (\theta, x) \mapsto (\theta + \omega, T_\theta(x)) \tag{1.1}$$

with irrational driving frequency ω. At most times, we will restrict to the case where the driving space $X = [a, b]$ is a compact interval and the *fibre maps* T_θ are all monotonically increasing on X. In this case we say T is a *qpf monotone interval map*. Some of the introductory examples will also be qpf circle homeomorphisms, but there the situation can often be reduced to the case of interval maps as well, for example when there exists a closed annulus which is mapped into itself.

Two notations which will be used frequently are the following: Given any set $A \subseteq \mathbb{T}^1 \times X$ and $\theta \in \mathbb{T}^1$, we let $A_\theta := \{x \in X \mid (\theta, x) \in A\}$. If $X = \mathbb{R}$ and $\varphi, \psi : \mathbb{T}^1 \to \mathbb{R}$ are two measurable functions, then we use the notation

$$[\psi, \varphi] \ := \ \{(\theta, x) \mid \psi(\theta) \leq x \leq \varphi(\theta)\} \tag{1.2}$$

similarly for (ψ, φ), $(\psi, \varphi]$, $[\psi, \varphi)$.

Due to the minimality of the irrational rotation on the base there are no fixed or periodic points for T, and one finds that the simplest invariant objects are invariant curves over the driving space (also invariant circles or invariant tori). More generally, a (T-)*invariant graph* is a measurable function $\varphi : \mathbb{T}^1 \to X$ which satisfies

$$T_\theta(\varphi(\theta)) \ = \ \varphi(\theta + \omega) \qquad \forall \theta \in \mathbb{T}^1 \ . \tag{1.3}$$

This equation implies that the point set $\Phi := \{(\theta, \varphi(\theta)) \mid \theta \in \mathbb{T}^1\}$ is forward invariant under T. As long as no ambiguities can arise, we will refer to Φ as an invariant graph as well.

There is a simple way of obtaining invariant graphs from compact invariant sets: Suppose $A \subseteq \mathbb{T}^1 \times X$ is T-invariant. Then

$$\varphi_A^+(\theta) \ := \ \sup\{x \in X \mid (\theta, x) \in A\} \tag{1.4}$$

defines an invariant graph (invariance following from the monotonicity of the fibre maps). Furthermore, the compactness of A implies that φ_A^+ is upper semi-continuous (see [**37**]). In a similar way we can define a lower semi-continuous graph φ_A^- by taking the infimum in (1.4). Particularly interesting is the case where $A = \bigcap_{n \in \mathbb{N}} T^n(\mathbb{T}^1 \times X)$ (the so-called global attractor, see [**34**]). Then we call φ_A^+ (φ_A^-) the *upper (lower) bounding graph of the system.*

There is also an intimate relation between invariant graphs and ergodic measures. On the one hand, to each invariant graph φ we can associate an invariant ergodic measure by

$$\mu_\varphi(A) \ := \ m(\pi_1(A \cap \Phi)) \ , \tag{1.5}$$

where m denotes the Lebesgue measure on $\mathbb{T}^1$ and π_1 is the projection to the first coordinate. On the other hand, if f is a qpf monotone interval maps then the converse is true as well: In this case, for each invariant ergodic measure μ there exists an invariant graph φ, such that $\mu = \mu_\varphi$ in the sense of (1.5). (This can be found in [**38**], Theorem 1.8.4 . Although the statement is formulated for continuous-time dynamical systems there, the proof literally stays the same.)

If all fibre maps are differentiable and we denote their derivatives by DT_θ, then the stability of an invariant graph φ is measured by its *Lyapunov exponent*

$$\lambda(\varphi) := \int_{\mathbb{T}^1} \log DT_\theta(\varphi(\theta))\, d\theta\ . \tag{1.6}$$

An invariant graph is called *stable* when its Lyapunov exponent is negative, *unstable* when it is positive and *neutral* when it is zero.

Obviously, even if its Lyapunov exponent is negative an invariant graph does not necessarily have to be continuous. This is exactly the case that has been the subject of so much interest:

DEFINITION 1.1 (Strange non-chaotic attractors and repellers). *A* ***strange non-chaotic attractor (SNA)*** *in a quasiperiodically forced system T is a T-invariant graph which has negative Lyapunov exponent and is not continuous. Similarly, a* ***strange non-chaotic repeller (SNR)*** *is a non-continuous T-invariant graph with positive Lyapunov exponent.*

This terminology, which was coined in theoretical physics, may need a little bit of explanation. For example, the point set corresponding to a non-continuous invariant graph is not a compact invariant set, which is usually required in the definition of 'attractor'. However, a SNA attracts and determines the behaviour of a set of initial conditions of positive Lebesgue measure (e.g. [**39**], Proposition 3.3), i.e. it carries a 'physical measure'. Moreover, it is easy to see that the essential closure[3] of a SNA is an attractor in the sense of Milnor [**3**]. 'Strange' just refers to the non-continuity and the resulting complicated structure of the graph. The term 'non-chaotic' is often motivated by the negative Lyapunov exponent in the above definition [**2**], but actually we prefer a slightly different point of view: At least in the case where the fibre maps are monotone interval maps or circle homeomorphisms, the topological entropy of a quasiperiodically forced system is always zero,[4] such that the system and its invariant objects should not be considered as 'chaotic'. This explains why we also speak of *strange non-chaotic repellers.* In fact, in invertible systems an attracting invariant graph becomes a repelling invariant graph for the inverse and vice versa, while the dynamics on them hardly changes. Thus, it seems reasonable to say that 'non-chaotic' should either apply to both or to none of these objects.

1.3. Examples of non-smooth saddle-node bifurcations

As mentioned, the crucial observation which starts our investigation here is the fact that the invariant circles in a non-smooth bifurcation do not approach each other arbitrarily. Instead, their behaviour follows a very distinctive pattern, which we call *exponential evolution of peaks.* In this section we present some simulations which demonstrate this phenomenom in the different parameter families mentioned in Section 1.1 . Although it seems difficult to give a precise mathematical definition

---

[3]The support of the measure μ_φ given by (1.5), where φ denotes the SNA. See also Section 3.1.

[4]For monotone interval maps this follows simply from the fact that every invariant ergodic measure is the projection of the Lebesgue measure on $\mathbb{T}^1$ onto an invariant graph, such that the dynamics are isomorphic in the measure-theoretic sense to the irrational rotation on the base. Therefore all measure-theoretic entropies are zero, and so is the topological entropy as their supremum. In the case of circle homeomorphisms, the same result can be derived from a statement by Bowen ([**40**], Theorem 17).

of this process, and we refrain from doing so here, this observation provides the necessary intuition and determines the strategy of the proofs for the rigorous results in the later chapters. (The same underlying idea can be found in [**33**] and [**35**].)

1.3.1. The arctan-family with additive forcing. Typical representatives of the class of systems we will study in the later sections are given by the family

$$(\theta, x) \mapsto \left(\theta + \omega, \frac{\arctan(\alpha x)}{\arctan(\alpha)} - \beta \cdot (1 - \sin(\pi\theta))\right). \tag{1.1}$$

As we will see later on, these maps provide a perfect model for the mechanism

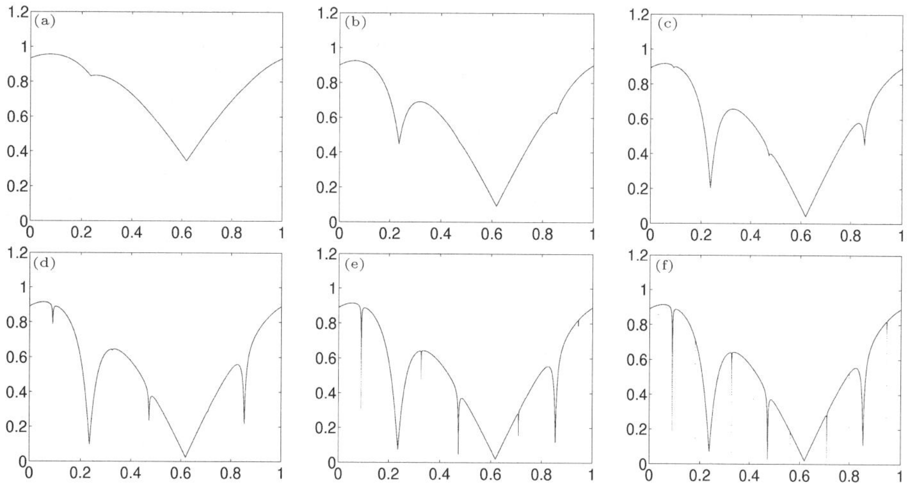

FIGURE 1.2. Upper bounding graphs in the parameter family given by (1.1) with ω the golden mean and $\alpha = 10$. The values for β are: (a) $\beta = 0.65$, (b) $\beta = 0.9$, (c) $\beta = 0.95$, (d) $\beta = 0.967$, (e) $\beta = 0.9708$, (f) $\beta = 0.9710325$.

which is responsible for the exponential evolution of peaks and the creation of SNA's in saddle-node bifurcations. The map $x \mapsto \frac{\arctan(\alpha x)}{\arctan(\alpha)}$ has three fixed points at 0 and ± 1, and for $\beta = 0$ these correspond to three (constant) invariant curves for (1.1). As the parameter β is increased, a saddle-node bifurcation between the two upper invariant curves takes place: Only the lower of the three curves persists, while the other two collide and cancel each other out. In fact, it will not be very hard to describe this bifurcation pattern in general (see Theorem 2.1), whereas proving that this bifurcation is indeed *'non-smooth'* will require a substantial amount of work.

Figure 1.2 shows the behaviour of the upper bounding graph as the parameter β is increased and reveals a very characteristic pattern. The overall shape of the curves hardly changes, apart from the fact that when the bifurcation is approached they have more and more *'peaks'* (as we will see there are infinitely many in the end, but most of them are too small to be seen). The point is that these peaks do not appear arbitrarily, but one after each other in a very ordered way: In (a), only the first peak is fully developed while the second just starts to appear. In (b) the second peak has grown out and a third one is just visible, in (c) and (d) the third one grows out and a fourth and fifth start to appear ... . Further, each peak is exactly the image of the preceding one, and the peaks become steeper and thinner at an exponential rate (which explains the term *'exponential evolution'* and the fact that the peaks soon become too thin to be detected numerically).

As far as simulations are concerned, the pictures obtained with smooth forcing functions in (1.1) instead of $(1-\sin(\pi\theta))$, which is only Lipschitz-continuous and decays linearly off its maximum at $\theta = 0$, show exactly the same behaviour. However, the rigorous results from the later sections only apply to this later type of forcing. In Section 1.4 we will discuss why this simplifies the proof of the non-smoothness of the bifurcation to some extent.

Finally, it should mentioned that the phenomenom we just described does not at all depend on any particular properties of the arcus tangent. Any strictly monotone and bounded map of the real line with the same qualitative features, which can vaguely described as being "s-shaped", can be used to replace the arcus tangent in the above definitions without changing the observed behaviour (e.g. $x \mapsto \tanh(x)$). If in addition this map has similar scaling properties as the arcus tangent, as for example $x \mapsto \frac{x}{1+|x|}$, then even the rigorous results we present in the later sections apply. We will not prove this in detail, but it will be evident that the arguments which we use in Section 2.4.1 to treat (1.1) can be easily adjusted to this end.

1.3.2. The Harper map. The Harper map with continuous *potential* $V : \mathbb{T}^1 \to \mathbb{R}$, *energy* E and *coupling constant* λ is given by

$$(\theta, x) \mapsto \left(\theta + \omega, \arctan\left(\frac{-1}{\tan(x) - E + \lambda V(\theta)}\right)\right) . \tag{1.2}$$

It is probably the most studied example, and the reason for this is the fact that its dynamics are intimately related to the spectral properties of discrete Schrödinger operators with quasiperiodic potential (the so-called almost-Mathieu operator in the case $V(\theta) = \cos(2\pi\theta)$). Before we turn to the simulations, we briefly want to discuss this relation and the arguments by which the existence of SNA in the Harper map is established in [**1**]. A more detailed discussion can be found in [**43**].

The map (1.2) describes the projective action of the $\mathrm{SL}(2,\mathbb{R})$-cocycle (or Schrödinger cocycle)

$$(\theta, v) \mapsto (\theta + \omega, A_{\lambda,E}(\theta) \cdot v) , \tag{1.3}$$

where

$$A_{\lambda,E}(\theta) = \begin{pmatrix} E - \lambda V(\theta) & -1 \\ 1 & 0 \end{pmatrix}$$

and $v = (v_1, v_2) \in \mathbb{R}^2$. This means that (1.2) can be derived from (1.3) by letting $x := \arctan(v_2/v_1)$. The Schrödinger cocycle in (1.3) is in turn associated to the almost-Mathieu operator

$$H_{\lambda,\theta} : \ell^2 \to \ell^2 , \ (H_{\lambda,\theta}u)_n = u_{n+1} + u_{n-1} + \lambda V(\theta + n\omega)u_n , \tag{1.4}$$

as each formal solution of the eigenvalue equation $H_{\lambda,\theta}u = Eu$ satisfies

$$\begin{pmatrix} u_{n+1} \\ u_n \end{pmatrix} = A_{\lambda,E}(\theta + n\omega) \cdot \begin{pmatrix} u_n \\ u_{n-1} \end{pmatrix} .$$

The existence of SNA for the Harper map is equivalent to *non-uniform hyperbolicity* of the cocycle (1.3) [**1, 43**], a concept which is of fundamental importance in this context.

In order to explain it, recall that a $SL(2,\mathbb{R})$-cocycle over an irrational rotation is a mapping $\mathbb{T}^1 \times \mathbb{R}^2 \to \mathbb{T}^1 \times \mathbb{R}^2$ of the form $(\theta, v) \mapsto (\theta + \omega, A(\theta) \cdot v)$, where $A : \mathbb{T}^1 \to SL(2,\mathbb{R})$ is a continuous function. The Lyapunov exponent of such a cocycle is given by

$$\lambda(\omega, A) = \lim_{n\to\infty} \frac{1}{n} \int_{\mathbb{T}^1} \log \|A_n(\theta)\| \, d\theta \,, \tag{1.5}$$

where $A_n(\theta) = A(\theta+(n-1)\omega)\circ\ldots\circ A(\theta)$. If $\lambda(\omega, A) > 0$, then Oseledets Multiplicative Ergodic Theorem implies the existence of an invariant splitting $\mathbb{R}^2 = \mathbb{W}^s_\theta \oplus \mathbb{W}^u_\theta$ (invariance meaning $A(\theta)(\mathbb{W}^i_\theta) = \mathbb{W}^i_{\theta+\omega}$ $(i = s, u)$), such that vectors in $\mathbb{W}^u_\theta$ are exponentially expanded and vectors in $\mathbb{W}^s_\theta$ are exponentially contracted with rate $\lambda(\omega, A)$ by the action of $A_n(\theta)$. The cocycle (ω, A) is called *uniformly hyperbolic* if the subspaces $\mathbb{W}^i_\theta$ depend continuously on θ. If they depend only measurably on θ, but not continuously, then the cocycle is called *non-uniformly hyperbolic.*

In order to see why the latter notion is equivalent to the existence of SNA, note that the invariant subspaces can be written as

$$\mathbb{W}^i_\theta = \mathbb{R} \cdot \begin{pmatrix} 1 \\ \tilde{\varphi}^i(\theta) \end{pmatrix}$$

with measurable functions $\tilde{\varphi}^i : \mathbb{T}^1 \to \mathbb{R} \cup \{\infty\}$, and it follows immediately that by $\varphi^i := \arctan(\tilde{\varphi}^i)$ we can define invariant graphs for the projective action of the cocycle (obtained by letting $x = \arctan(v_2/v_1)$ as above). Moreover, it is not difficult to show that $\lambda(\varphi^s) = 2\lambda(\omega, A)$ and $\lambda(\varphi^u) = -2\lambda(\omega, A)$ in this case,[5] and conversely the existence of invariant graphs with non-zero Lyapunov exponent implies the existence of an invariant splitting with the mentioned properties. As the graphs φ^i depend continuously on θ if and only if this is true for the subspaces $\mathbb{W}^i_\theta$, we obtain the claimed equivalence.

The crucial observation which was made by Herman is the fact that, using a result from sub-harmonic analysis, lower bounds on the Lyapunov exponent can be obtained for suitable choices of the $SL(2,\mathbb{R})$-valued function A. In the case of (1.3) with potential $V(\theta) = \cos(2\pi\theta)$, this bound is $\lambda(\omega, A_{\lambda,E}) \geq \max\{0, \log(|\lambda|/2)\}$ [**1**, Section 4.7]. Consequently, if $|\lambda| > 2$ then the Lyapunov exponent of $(\omega, A_{\lambda,E})$ will be strictly positive for all values of E. On the other hand, it is well-known that there cannot be a continuous splitting for all $E \in \mathbb{R}$, and consequently for some E the respective cocycle has to be non-uniformly hyperbolic.

---

[5] In the case of the Harper map, the crucial computation is the following: Fix $\theta \in \mathbb{T}^1$ and $v \in \mathbb{R}^2 \setminus \{0\}$ and define vectors v^n by $v^0 := v$ and $v^{n+1} := A(\theta + n\omega) \cdot v^n$. Further, let $\theta_n := \theta + n\omega \bmod 1$ and $x_n := \arctan(v_2^n/v_1^n)$, and denote the Harper map (1.2) by T. Then,

$$\begin{aligned} DT_{\theta_k}(x_k) &= \frac{1}{1+(\tan(x_k) - E + \lambda V(\theta_k))^{-2}} \cdot \frac{1+\tan(x_k)^2}{(\tan(x_k) - E + \lambda V(\theta_k))^2} \\ &= \frac{1+\tan(x_k)^2}{1+(\tan(x_k) - E + \lambda V(\theta_k))^2} = \frac{1+\tan(x_k)^2}{1+\tan(x_{k+1})^{-2}} = \frac{\|v^k\|^2}{\|v^{k+1}\|^2} \,, \end{aligned}$$

where we used $v_2^{k+1} = v_1^k$ in the last step. Consequently, we obtain

$$DT^n_\theta(x_0) = \prod_{k=0}^{n-1} DT_{\theta_k}(x_k) = \frac{\|v^0\|^2}{\|v^n\|^2} \,,$$

and this establishes the asserted relation between the different Lyapunov exponents. The case of a general $SL(2,\mathbb{R})$-cocycle can be treated in more or less the same way.

The simplest way to see this is probably to consider the rotation number. Suppose $\omega \in \mathbb{T}^1 \setminus \mathbb{Q}$ and $\lambda > 2$ are fixed. Then (1.2) defines a skew-product map T_E on the two-torus, and for such maps a fibred rotation number $\rho(T_E)$ can be defined, much in the way this is done for homeomorphisms of the circle. The dependence of $\rho(T_E)$ on E is continuous [**1**, Section 5], and further it is easy to see that the existence of continuous invariant graphs forces the fibred rotation number to be rationally related to ω, more precisely to take values in the module $\mathcal{M}_\omega := \{\frac{k}{q}\omega \bmod 1 \mid k \in \mathbb{Z}, q \in \mathbb{N}\}$ (compare [**1**, Section 5.17]). However, if E runs through the real line from $-\infty$ to ∞, then the rotation number $\rho(T)$ runs exactly once around the circle [**1**, Section 4.17(b)]. For all $E \in \mathbb{R}$ with $\rho(T_E) \notin \mathcal{M}_E$, the existence of a SNA in (1.2) follows. Refined results can be obtained by using the fact that the invariant splitting cannot be continuous whenever E belongs to the spectrum of the almost-Mathieu operator. This is discussed in detail in [**43**]. In particular, it allows to use lower bounds on the measure of the spectrum to establish the existence of SNA for a set of positive measure in parameter space.

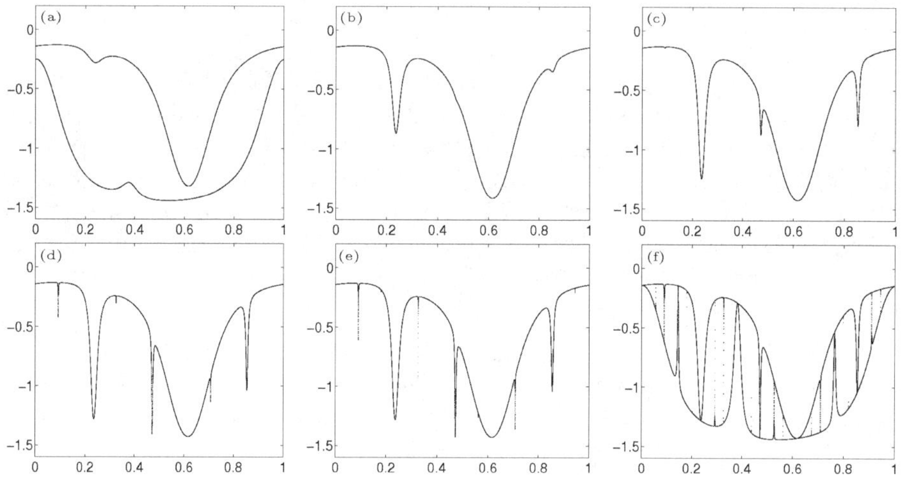

FIGURE 1.3. The stable invariant curves for the projected Harper map given by (1.6) with ω the golden mean, $\lambda = 4$ and different values for E. (a) At $E = 4.4$ the first peak is clearly visible, while the second just starts to appear. The repeller is close, but still a certain distance away. (b) At $E = 4.3$ the second peak has grown and the third starts to appear. This pattern continues, and more and more peaks can be seen in pictures (c) $E = 4.289$, (d) $E = 4.28822$ and (e) $E = 4.288208$. (f) finally shows attractor and repeller for $E = 4.288207478$ just prior to collision.

For the simulations presented here we use a reflection of (1.2) w.r.t. the θ-axis,

$$(1.6) \qquad (\theta, x) \mapsto \left(\theta + \omega, \arctan\left(\frac{1}{\tan(-x) - E + \lambda V(\theta)}\right)\right),$$

as this makes it easier to compare the pictures with the other examples. The potential function which is used is $V(\theta) = \cos(2\pi\theta)$. Later, in Section 2.4.2, we have to make a different choice in order to obtain rigorous results with the methods presented here.

As described by Herman in [**1**, Section 4.14], when the parameter E approaches the spectrum of the almost-Mathieu operator from above, a stable and an unstable invariant circle collide in a saddle-node bifurcation. Even if the rigorous arguments used by Herman [**1**] are very specific for cocycles (as described above), the process seems to be the same as in the arctan-family before: Figure 1.3 shows the behaviour of the attractor before it collides with the repeller (the latter is only depicted in Fig. 1.3(a) and (f)). The pattern is already familiar, the exponential evolution of peaks can be seen quite clearly again.

Based on this observation, Bjerklöv recently addressed a problem raised by Herman [**1**, Section 4.14] about the structure of the minimal set which is created in this bifurcation. Upon their collision, the stable and unstable invariant circles are replaced by an upper, respectively lower semi-continuous invariant graph. The region between the two graphs is a compact and invariant set, but it is not at all obvious whether this set is also minimal and coincides with the topological closures of the two graphs. In [**33**] Bjerklöv gives a positive answer to this question, provided the rotation number ω on the base is Diophantine and the parameter λ is sufficiently large. As his approach is purely dynamical and does not depend on any particular properties of cocycles, it should be possible to apply it to more general systems. This might allow to prove the existence of SNA and to describe their structure, in the above sense, at the same time.

1.3.3. The quasiperiodically forced Arnold circle map. The most obvious physical motivation for studying qpf systems are probably oscillators which are forced at two or more incommensurate frequencies. If these are modelled by differential equations, the Poincaré maps will be of the form (1.1). The qpf Arnold circle map, given by

$$(\theta, x) \mapsto \left(\theta + \omega, x + \tau + \frac{\alpha}{2\pi}\sin(2\pi x) + \beta\sin(2\pi\theta)\right) \tag{1.7}$$

with real parameters α, τ and β, is often studied as a basic example (see [**9**]). There are several interesting phenomena which can be found in this family, such as different bifurcation patterns, mode-locking or the transition to chaos as the map becomes non-invertible [**9, 12, 5**]. Similar to the unforced Arnold circle map [**41, 42**], there exist so-called Arnold tongues – regions in the parameter space on which the rotation number stays constant. The reason for this is usually the existence of (at least) one stable invariant circle inside of the tongue. On the boundaries of the tongue this attractor collides with an unstable invariant circle in a saddle-node bifurcation (see [**5, 14**] or [**17**] for a more detailed discussion and numerical results).

For our purpose it is convenient to study only those bifurcations which take place on the boundary of the Arnold tongue with rotation number zero. In order to do so, we fix the parameters $\alpha \in [0,1]$ and $\beta > 0$, thus obtaining a one-parameter family depending on τ. As long as β is not too large, there exist a stable and an unstable invariant curve at $\tau = 0$. Increasing or decreasing τ leads to the disappearance of the two curves after their collision in a saddle-node bifurcation. When α is close enough to 1 (where the map becomes non-invertible) this bifurcation seems to be non-smooth [**5**]. The problem here is the fact that the curves are already

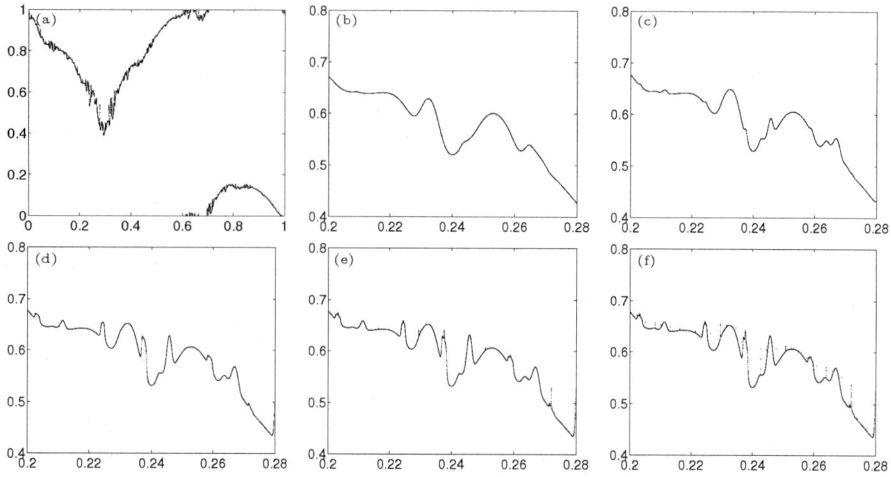

FIGURE 1.4. Pictures obtained from the qpf Arnold circle map (1.7) with $\alpha = 0.99$ and $\beta = 0.6$, ω is the golden mean. (a) shows the attracting invariant curve for $\tau = 0.3373547962$. As the exponential evolution of peaks takes place on a rather microscopic level, it is difficult to recognise any details. Therefore, the other pictures show the attractors only over the interval $[0.2, 0.28]$. The τ-values are (b) $\tau = 0.337$, (c) $\tau = 0.3373$, (d) $\tau = 0.3373547$, (e) $\tau = 0.33735479$, (f) $\tau = 0.337357962$.

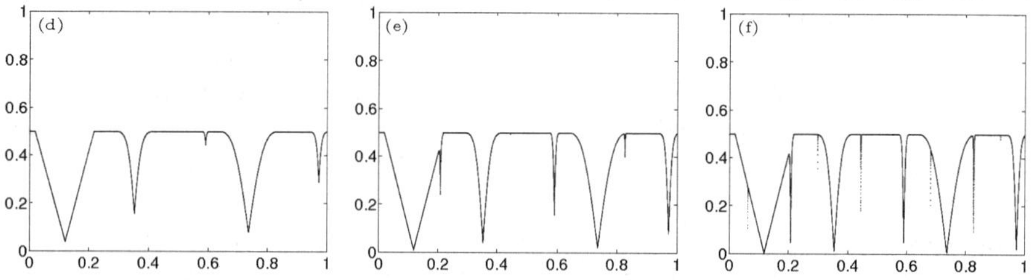

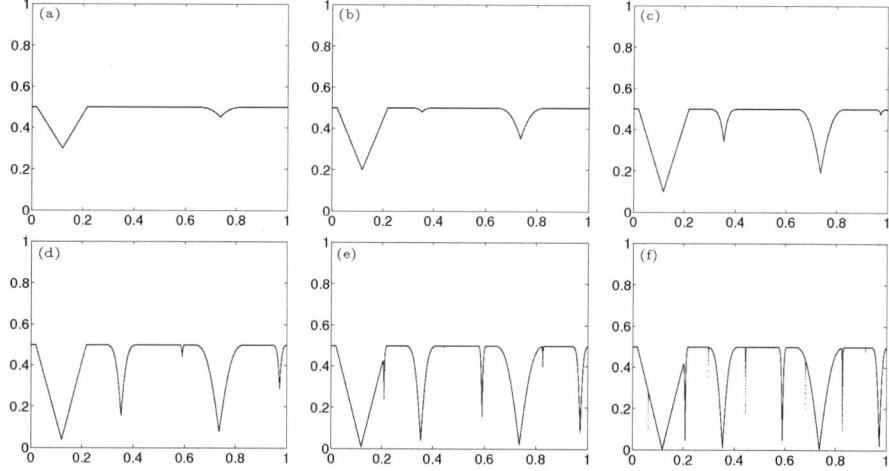

FIGURE 1.5. The stable invariant curves in the system $(\theta, x) \mapsto (\theta + \omega, x + \tau + \frac{\alpha}{2\pi}\sin(2\pi\theta) \; -\beta \cdot \max\{0, 1 - 10 \cdot d(\theta, \frac{1}{2})\})$. This time the parameters $\alpha = 0.99$ and $\tau = 0$ are fixed, while β varies: (a) $\beta = -0.2$, (b) $\beta = -0.3$, (c) $\beta = -0.4$, (d) $\beta = -0.45$, (e) $\beta = -0.49$, (f) $\beta = -0.497$. Again, ω is the golden mean. The exponential evolution of peaks is clearly visible.

extremely 'wrinkled' before the exponential evolution of peaks really starts. Therefore, it is hard to recognise any details in the global picture (see Figure 1.4(a)). This becomes different if we 'zoom in' and only look at the curves over a small interval. On this microscopic level, we discover the more or less the same behaviour as before (Figure 1.4(b)–(f)). Of course, this time we can not really determine the order in which the peaks are generated, as we only see those peaks which lie in our small interval. But we clearly see that more and more peaks appear, and those appearing at a later time are smaller and steeper than those before.

On the other hand, we can also use a more '*peak-shaped*' forcing function instead of the sine. In this case, the pictures we obtain look exactly the same as the ones from the arctan-family above (see Figure 1.5(a)-(f)). This effect will be discussed in more detail in Section 1.4 . Nevertheless, we should mention that, in contrast to the two preceding examples, we do not provide any rigorous results on the qpf Arnold circle map (see also Section 2.4.3 for a discussion).

1.3.4. Pinched skew products. As for the Harper map, we refer to the original literature [**2, 4**] for a more detailed discussion of these systems. Here, we will just have a look at the map

$$(\theta, x) \mapsto (\theta + \omega, \tanh(\alpha x) \cdot \sin(\pi\theta)) \;, \tag{1.8}$$

with real positive parameter α, which is a typical representative of this class of systems. Note that due to the multiplicative nature of the forcing, the 0-line is *a priori* invariant, and due to the zero of the sine function there is one fibre which is mapped to a single point (hence 'pinched'). These are the essential features that are needed to prove the existence of SNA in pinched skew products (see [**4, 34**]).

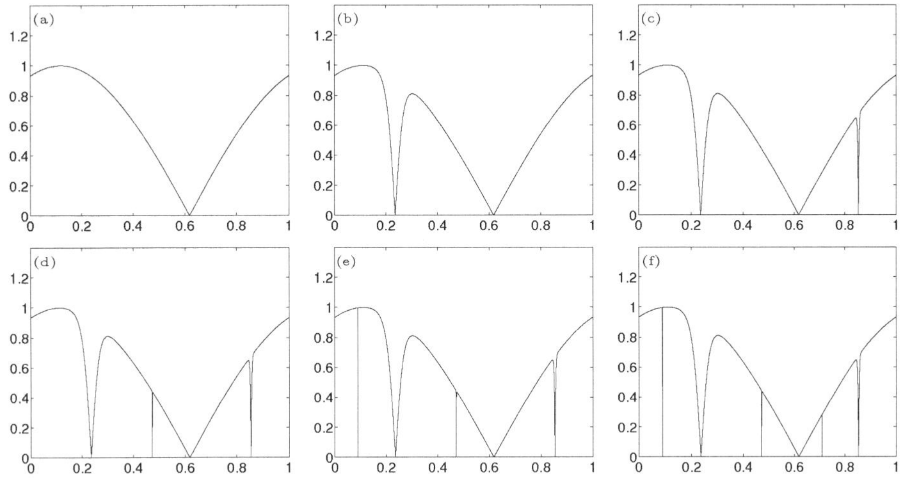

FIGURE 1.6. The first six iterates of the upper boundary line for the pinched skew product given by (1.8) with ω the golden mean and $\alpha = 10$. In each step of the iteration one more peak appears, while apart from that the curves seem to stay the same. Further, the peaks become steeper and thinner at an exponential rate.

Figure 1.6 differs from the preceding ones insofar as it does not show a sequence of invariant graphs as the systems parameters are varied, but the first images of a constant line that is iterated with a fixed map. Nevertheless, the behaviour is very much the same as before. The exponential evolution of peaks can followed even easier here, as this time each iterate produces exactly one further peak.

For Pinched skew products this process was quantified [**35**] in order to describe the structure of the SNA's in more detail. The question addressed there is basically the same as the one studied by Bjerklöv in [**33**], and the result is similar: The SNA, which is an upper semi-continuous invariant graph above the 0-line in this situation, lies dense in the region below itself and above the 0-line, provided the rotation number ω on the base is Diophantine and the parameter α is large enough.

1.3.5. Non-smooth pitchfork bifurcations. Compared to saddle-nodes, pitchfork bifurcations are degenerate. Usually they only occur if the system has some inherent symmetry that forces three invariant circles to collide exactly at the same time. Nevertheless, they have been described in the literature about SNA's quite often (e.g. [**5**],[**16**]). The reason for this is the fact that unlike in saddle-node bifurcations, where the SNA's only occur at one single parameter, SNA's which are created in pitchfork bifurcations seem to persist over a small parameter interval. In addition, the transition from continuous to non-continuous invariant graphs at the collision point is much more distinct, as the SNA which is created seems to trace out a picture of both stable invariant curves just prior to the collision (see Figure 1.7).

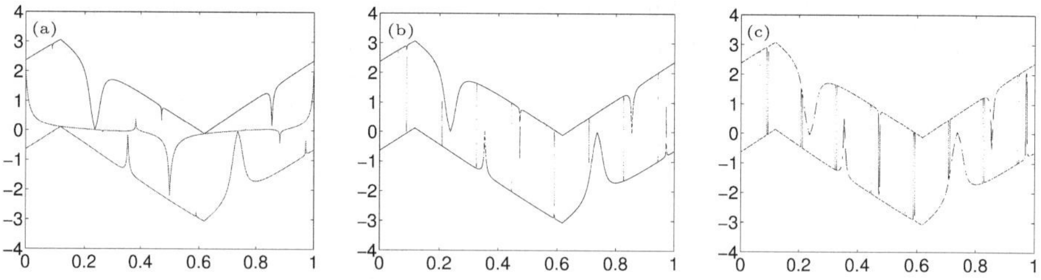

FIGURE 1.7. A pitchfork bifurcation in the parameter family (1.9). (a) shows the upper and lower bounding graphs just prior to the collision. Note that here the two objects are still distinct, and three different trajectories (a backwards trajectory for the repeller) are plotted to produce this picture. In contrast to this, (b) and (c) only show one single trajectory. There still exist two distinct SNA's, but these are interwoven in such a way that they cannot be distinguished anymore. Each of them seems to trace out a picture of both attractors before collision. The parameter values are $\alpha = 10$ and (a) $\tau = 1.64$, (b) $\tau = 1.645$ and (c) $\tau = 1.66$. ω is the golden mean.

We were not able to give a rigorous proof for this stabilising effect, or any other details of a non-smooth pitchfork bifurcation. However, by a slight modification of the methods used for the non-smooth saddle-node bifurcation, we can at least prove the existence of SNA's in systems with the mentioned inherent symmetry (see Theorem 2.10 and Section 2.4.4). For suitable parameters these systems have two SNA's which are symmetric to each other and enclose a self-symmetric SNR, and the three objects are interwoven in such a way that they all have the same (essential) topological closure. As an example, we consider the parameter family

$$(\theta, x) \mapsto \arctan(\alpha x) - \beta \cdot (1 - 4d(\theta, 0)) \ . \tag{1.9}$$

For Diophantine ω and sufficiently large α we will obtain the existence of a SNA-SNR triple as described above for at least one suitable parameter $\beta(\alpha)$.

1.4. The mechanism: Exponential evolution of peaks

In the following, we will try to give a simple heuristic explanation for the mechanism which is responsible for the exponential evolution of peaks. Generally, one could say that it consists of a subtle interplay of an *'expanding region'* $\mathcal{E}$ and a *'contracting region'* $\mathcal{C}$, which communicate with each other only via a small *'critical region'* $\mathcal{S}$. In order to give meaning to this, we concentrate first on the arctan-family given by (1.1).

If we restrict to $\alpha \geq \tan(1)$ and $\beta \leq \pi$ in (1.1), then we can choose $X = \left[-\frac{3}{2}\pi, \frac{3}{2}\pi\right]$ as the driven space, because in this case $\mathbb{T}^1 \times \left[-\frac{3}{2}\pi, \frac{3}{2}\pi\right]$ is always mapped into itself. Further, we fix α sufficiently large, such that the map $F : x \mapsto \arctan(\alpha x)$ has three fixed points $x^- < 0 < x^+$. As 0 will be repelling and x^+ attracting, we can choose a small interval I_e around 0 which is expanded and an interval I_c around x^+ which is contracted, and define the expanding and contraction regions as $\mathcal{E} := \mathbb{T}^1 \times I_e$ and $\mathcal{C} := \mathbb{T}^1 \times I_c$ (see Figure 1.8). Of course, there exists a second contracting region $\mathcal{C}^-$, corresponding to x^-, but this does not take part in the bifurcation: Due to the one-sided nature of the forcing, $\mathcal{C}^-$ is always a trapping region, independent of the parameter β. Thus there always exists a stable invariant circle inside of $\mathcal{C}^-$, and the saddle-node bifurcation only takes place between the two invariant circles above.

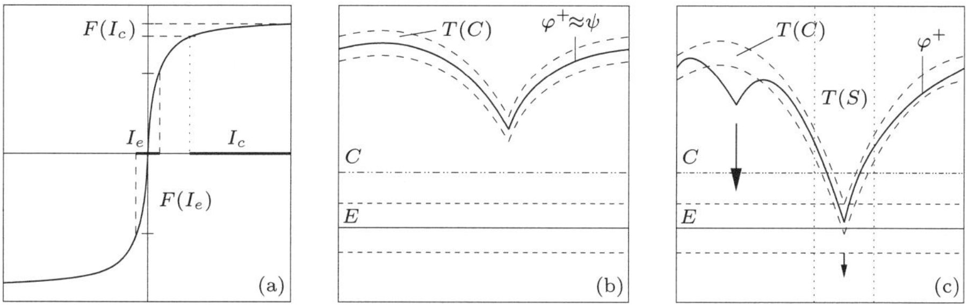

FIGURE 1.8. As the fibre maps are expanding and contracting on I_e and I_c, respectively, T will be expanding in the vertical direction on $\mathcal{E} = \mathbb{T}^1 \times I_e$ and contracting on $\mathcal{C} = \mathbb{T}^1 \times I_c$. (b) As long as β is not too large, $\mathcal{C}$ is mapped into itself. Thus, there exists a stable invariant circle inside of $T(\mathcal{C})$ (in fact, as a point set this circle coincides with $\bigcap_{n\in\mathbb{N}} T^n(\mathcal{C})$), which has approximately the shape of the forcing function. (c) When the first peak enters the expanding region it induces a second peak, which moves faster than the first one by the expansion factor in $\mathcal{E}$. The first peak is generated in the critical region $\mathcal{S}$, where the forcing achieves its maximum. Therefore, it is located in $T(\mathcal{S})$.

By the choice of the intervals, the fibre maps T_θ are contracting on I_c and expanding on I_e. Further, as long as β is small there holds

$$T_\theta(I_c) \subseteq I_c \quad \text{and} \quad I_e \subseteq T_\theta(I_e) \tag{1.1}$$

for all $\theta \in \mathbb{T}^1$. Consequently,

$$T(\mathcal{C}) \subseteq \mathcal{C} \quad \text{and} \quad \mathcal{E} \subseteq T(\mathcal{E}) \ . \tag{1.2}$$

This means that $\mathcal{C}$ and $\mathcal{E}$ cannot interact, and there will be exactly one invariant circle (stable and unstable, respectively) in each of the two regions. However, when β is increased and approaches the bifurcation point, (1.2) does not hold anymore. Nevertheless, the relation (1.1) will still be true for 'most' θ, namely whenever the forcing function $(1-\sin(\pi\theta))$ in (1.1) is not close to its maximum (see Figure 1.8(c)). Thus, even when $\mathcal{E}$ and $\mathcal{C}$ start to interact, they will only do so in a vertical strip $\mathcal{S} := W \times X$, where $W \subseteq \mathbb{T}^1$ is a small interval around 0.

This strip $\mathcal{S}$ is the '*critical region*' we referred to above and in which the first peak is generated: As long as $T(\mathcal{C}) \subseteq C$, the upper bounding graph will be contained in $T(\mathcal{C})$. But this set is just a very small strip around the first iterate of the line $\mathbb{T}^1 \times \{x^+\}$, which is a curve ψ given by

$$\psi(\theta) := x^+ - \beta \cdot (1 - \sin(\pi(\theta - \omega))) \tag{1.3}$$

(see Figure 1.8(b)). Consequently, the upper bounding graph φ^+ will have approximately the same shape as ψ, which means that it has a first peak centred around ω, i.e. in $T(\mathcal{S})$. From that point on, the further behaviour is explained quite easily. As soon as the first peak enters the expanding region, its movement will be amplified due to the strong expansion in $\mathcal{E}$. Thus a second peak will be generated at 2ω mod 1. It will be steeper than the first one, and when β is increased it also grows faster by a factor which is more or less the expansion factor inside $\mathcal{E}$. As soon as the second peak is large enough to enter the expanding region, it generates a third one, which in turn induces a fourth and so on ... .

The picture we have drawn so far already gives a first idea about what happens, although converting it into a rigorous proof for the existence of SNA will still require a substantial amount of work. As we will see, it is not too hard to give a good quantitative description of the behaviour of the peaks up to a certain point, namely as long as the peaks do enter the critical region (corresponding to the returns of the underlying rotation to the interval W). But as soon as this happens, things will start to become difficult. However, by assuming that the rotation number ω satisfies a Diophantine condition we can ensure that such returns are not too frequent, and that very close returns do not happen too soon. This will be sufficient to ensure that the exponential evolution of peaks also carries on afterwards.

In principle, the mechanism is not different in the other parameter families discussed in the last section. For the Harper map given by (1.6), Figure 1.9(a) shows the graph of a projected Möbius-transformation $x \mapsto \arctan(\frac{1}{\tan(-x)-c})$ for large c. As long as $E \gg \lambda$, the fibre maps will all have approximately this shape. As we can see, there will be a repelling fixed point slightly above $-\frac{\pi}{2}$ and an attracting one slightly below 0. This means that if we choose I_e and I_c to be sufficiently small intervals around these fixed points, then we have uniform expansion on $\mathcal{E}$, uniform contraction on $\mathcal{C}$ and (1.1) will be satisfied. When $E \approx \lambda$, this will still be true on most fibres. Only where the potential $\cos(2\pi\theta)$ is close to its maximum at $\theta = 0$,

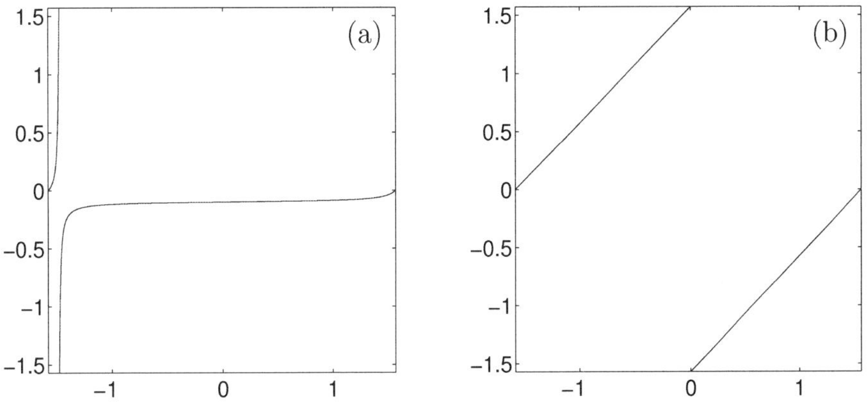

FIGURE 1.9. Graphs of the projected Möbius-transformations $x \mapsto \arctan\left(\frac{1}{\tan(-x)-10}\right)$ in (a) and $x \mapsto \arctan\left(\frac{1}{\tan(-x)}\right)$.

the picture changes (Figure 1.9(b)). Here $-\frac{\pi}{2} \in I_c$ is mapped close to 0, which means again that the expanding and contracting region start to interact and a first peak is produced. (Thus, the critical region $\mathcal{S}$ is again a vertical strip around 0.) As before, this peak is amplified as soon as it enters the expanding region $\mathcal{E}$ and thus induces all others.

In some sense, the situation for the qpf Arnold circle map is even more similar to the case of the arctan-family, as the forcing is additive again and the fibre maps are clearly s-shaped as before. However, the difference is the fact that while the derivative at the stable fixed point indeed vanishes, such that the contraction becomes arbitrarily strong, the maximal expansion factor is at most 2 (at least in the realm of invertibility $\alpha \leq 1$). This explains why the resulting pictures in Figure 1.4 are much less clear. Roughly speaking, in combination with the limited expansion the peak of the forcing function $\theta \mapsto \sin(\pi\theta)$ is just 'too blunt' to trigger the exponential evolution of peaks as easily as before. When it finally does take place - as the simulations in Figure 1.4 suggest - the graphs are already too 'wrinkled' to give a good picture. But of course, if the shape of the forcing function is a second factor that decides whether the exponential evolution of peaks takes place, then we can also trigger this pattern by choosing one with a very sharp peak. This is exactly what happened in Figure 1.5.

Finally, for Pinched skew products we refer to [**35**] for a more detailed discussion.

REMARK 1.2. The preceding discussion gives a basic understanding of how SNA's are created in the above examples. Although it might be very rudimentary, it can already be used to anticipate a number of observations. Without trying to make things very precise, we want to mention a few:

(a) First, it is not hard to guess in which parameter range the expanding and contracting regions start to interact and the torus collision takes place in the above families, e.g. $E \approx \beta$ for the Harper map or $\beta \approx \frac{\pi}{2}$ for the arctan-family.

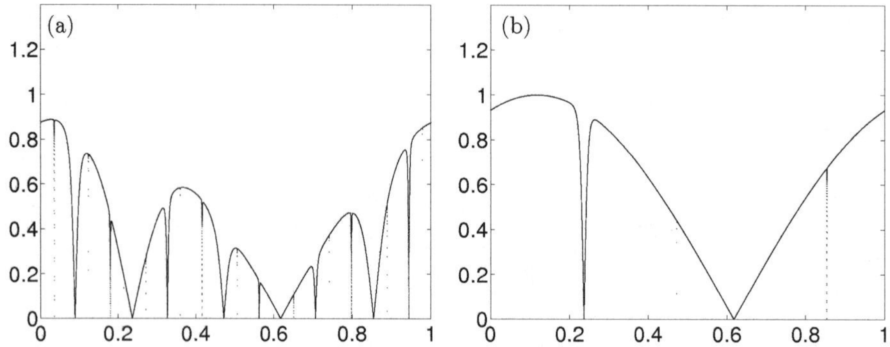

FIGURE 1.10. Upper bounding graphs in the pinched systems given by (1.8). ω is the golden mean, the parameter values are (a) $\alpha = 3$ and (b) $\alpha = 32$. In (b), where the expansion is stronger, there seems to be less structure in comparison to (a). However, this is not a qualitative difference, but can be easily explained by the exponential evolution of peaks. If the expansion is stronger, the peaks of higher order are just not visible anymore, such that only the first few peaks can be seen.

(b) Another phenomenom which can be explained is the following: The stronger the expansion and contraction are, i.e. the larger the respective parameter is chosen, the less 'structure' can be seen in the pictures (see Figure 1.10). However, obviously this 'structure' corresponds exactly to the peaks which are generated. These can only be detected numerically as long as they do not become too small, but of course this happens faster if the expansion and contraction are stronger. Figure 1.10 shows this effect for pinched systems, but it can be observed similarly in all the examples we treated. In particular, it is also present in the qpf Arnold circle map (1.7), which indicates again that the mechanism there is not different from the other examples.

(c) As already mentioned, the exponential evolution of peaks is easier to trigger if the forcing function has a very distinctive and sharp peak. Figures 1.4 and 1.5 illustrate this in the context of the qpf Arnold circle map.

(d) In [**20**], the authors study (amongst other things) the parameter-dependence of the minimal distance Δ_β between the stable and instable invariant curve in a non-smooth bifurcation. Their situation is slightly different to the one considered here, since the dynamics take place on a torus and the attractor touches the repeller from above and below at the same time. Nevertheless, the pictures indicate that a process similar to the one described above takes place. The observation which was made by the authors is that the asymptotic dependence of Δ_β on $|\beta - \beta_c|$ seems to be a power law with exponent 1 as $\beta \to \beta_c$, i.e. $\Delta_\beta \sim |\beta - \beta_c|$ (where β_c is the bifurcation parameter). Furthermore, this exponent seems to be universal for a certain class of models.

At least in the situations we discussed, e.g. for (1.1), the exponential evolution of peaks offers a reasonable explanation for such a scaling behaviour: Since all peaks of the attractor have to touch the repeller at the same time and, according to our heuristics, all further peaks move much faster than the first one, it is the latter which determines the minimal distance of the two curves. However, as this first peak has approximately the shape of the forcing function (see (1.3)), the position of its tip depends linearly on β.

Admittedly, some of the above remarks remain rather speculative unless they are confirmed either by careful numerical studies or rigorous proofs. Nevertheless, what we want to point out is that the mechanism we described offers at least an heuristic explanation for a number of observations which have sometimes been found to be puzzling or ever confusing. Further, an intuitive understanding of the process should make it easier to come up with reasonable conjectures, which can then (in the better case) either be proved or at least be confirmed numerically. As already mentioned, the issue we want to concentrate on in this article is a rigorous proof for the existence of SNA.

Concerning the latter, the main problem we will encounter is that we do not a priori know where the tips of the peaks are located. If there is any chance of rigorously describing the exponential evolution of peaks in a quantitative way, they must be located in the expanding region at least most of the times. Otherwise, there would be no plausible mechanism which forces the peaks to become steeper and steeper. But the horizontal position is not the only problem. When we use a forcing function with a quadratic maximum, then we do not even know the exact vertical position: If the tip of one peak is on the fibre θ, then the tip of the next will be close to $\theta + \omega$, but it may be slightly shifted due to the influence of the forcing. In order to explain this, suppose that a upper bounding graph φ^+ of T is differentiable and has a local minimum at θ_0. The derivative of φ at $\theta_0 + \omega$ is then given by

$$\varphi'(\theta_0 + \omega) \;=\; \frac{\partial}{\partial_\theta}(\pi_2 \circ T)(\theta_0, \varphi(\theta_0)) + DF_\theta(\varphi(\theta)) \cdot \varphi'(\theta_0) \;=\; -\beta g'(\theta_0)\ . \tag{1.4}$$

(Here we suppose that T has fibre maps of the form $T_\theta = F(x) - \beta g(\theta)$ as in (1.1).) Consequently, if $g'(\theta_0) \neq 0$, then $\theta_0 + \omega$ is not a local minimum.

This becomes different if the local minima, which we call 'peaks', are sufficiently 'sharp and steep'. By this, we mean that that both $\lim_{\theta \nearrow \theta_0} -\varphi'(\theta)$ and $\lim_{\theta \searrow \theta_0} \varphi'(\theta)$ are greater than a sufficiently large constant M (depending on the $\mathcal{C}^1$-norms of F and g). Then it can easily be seen from (1.4) that $\theta_0 + \omega$ will be a local minimum as well. If in addition $(\theta_0, \varphi(\theta_0))$ is located in the expanding region and the expansion constant is sufficiently large, then the peak at $\theta_0 + \omega$ will again be sufficiently sharp and steep (in the above sense).

Our claim is now that we can produce such sharp peaks by choosing a forcing function that, like $1 - \sin(\pi\theta)$, is only Lipschitz-continuous at its maximum and decays linearly in a neighbourhood. At least for the first peak this is plausible, since we have argued above that at the onset of the exponential evolution of peaks the invariant graph has approximately the shape of the forcing function (see the discussion around (1.3)). For all further peaks we can expect the same, provided that the exponential evolution of peaks is really caused by the mechanism described above, because then the tips of the peaks are located in the expanding region (at least most of the time).

However, we will not give a rigorous proof for this claim, since this would require to describe the global structure of the invariant graphs. In fact, we argue that it is exactly this 'localisation' of the tips of the peaks which helps to overcome the need for such a global description (which would probably be much more complicated on a technical level). In order to understand this, note that (in case our claim holds), the tips of the peaks just correspond to a single orbit, since then one minimum is mapped to another. Further, as mentioned, we expect that this orbit spends most of the time in the expanding region, and in fact this will already turn out to be sufficient to prove the existence of a SNA: In this case there exists an orbit on the upper bounding graph which has a positive vertical Lyapunov exponent, and this is not compatible with the continuity of the upper bounding graph (the Lyapunov exponent of the upper bounding graph is always non-positive, e.g. Lemma 3.5 in [**39**], and due to uniform convergence of the ergodic limits this is true for any of its points).

However, during the proof we will obtain even more information about this particular orbit: It does not only have a positive Lyapunov exponent forwards, but also backwards in time. Thus, concerning it Lyapunov exponents the orbit behaves as if it was moving from a sink to a source (and referring to this we will call it a *'sink-source-orbit'*). As it will turn out, it is contained in the intersection of the SNA and the SNR. The existence of such atypical orbits is also well-known for the Harper map, where it is equivalent to the existence of exponentially decaying eigenfunctions for the associated Schrödinger operators[6] and indicates an intersection of the stable and unstable subspaces of the matrix cocycle (see [**43**] for a more detailed discussion).

Summarising we can say that the 'sharp' peak makes it possible to concentrate on a single orbit instead of a whole sequence of graphs, and the information about this orbit will already be sufficient to establish the existence of a SNA. The fact that the construction in the proof of our main results (Theorems 2.7 and 2.10), which is based of this idea, works fine can be seen as an indirect 'proof' of the claim we made in the above discussion.

[6]Suppose $u \in \ell^2$ is a non-zero solution of the eigenvalue equation (1.4) and let $x_n = \arctan(u_{n-1}/u_n)$ (see Section 1.3.2). Further, denote the Harper map given by (1.2) by T. Then, using the formula derived in Footnote 5, we obtain that

$$DT_\theta^n(x_0) \quad = \quad \frac{u_0^2 + u_{-1}^2}{u_n^2 + u_{n-1}^2} \ .$$

(Note that $u_{n-1} = u_n = 0$ is not possible, as otherwise $u = 0$.) Consequently, sink-source-orbits correspond to exponentially decaying eigenfunctions. The existence of such 'localised' eigenfunctions for the almost-Mathieu operator was shown by Jitomirskaya in [**44**] (so-called *Anderson localisation*).

CHAPTER 2

Statement of the main results and applications

In this section we state and discuss the main results of this article and their application to the examples from the introduction. The proofs are postponed until the later sections, unless they can be given in a few lines. In particular, this concerns the construction of sink-source-orbits, which is carried out in Sections 4 to 7.

Before we turn to results on the non-smoothness of bifurcations in Section 2.3, we provide a general setting in which saddle-node bifurcations in qpf interval maps take place (Section 2.1), and introduce sink-source-orbits as a criterium for the existence of SNA (Section 2.2).

2.1. A general setting for saddle-node bifurcations in qpf interval maps

Obviously, before we can study the non-smoothness of saddle-node bifurcations, we have to provide a setting in which such bifurcations occur (smooth or non-smooth). In order to do so, we will consider parameter families of maps $T = T_\beta$ which are given by

$$T_\beta(\theta, x) \ := \ (\theta + \omega, F(x) - \beta \cdot g(\theta)) \ , \tag{2.1}$$

where we suppose that, given a constant $C > 0$, the functions F and g satisfy the following assumptions:

(2.2) $g : \mathbb{T}^1 \to [0, 1]$ is continuous and takes the value 1 at least once;

(2.3) $F : [-2C, 2C] \to [-C, C]$ is continuously differentiable with $F' > 0$;

(2.4) F has exactly three fixed points $x_- < 0, 0$ and $x_+ > 0$.

Note that if we restrict to parameters $\beta \in [0, C]$, then we can choose $X = [-2C, 2C]$ as the driven space, because then $\mathbb{T}^1 \times X$ is always mapped into itself. Of course, this choice is somewhat arbitrary, the only thing which is important is to fix some driven space X independent of the parameter β. We also remark that the in the situations we will consider later, F is usually a bounded function which is defined on the whole real line. In this case, we will only consider its restriction $F_{|[-2C,2C]}$, where C is any constant larger that $\sup_{x \in \mathbb{R}} |F(x)|$. This has the advantage that we obtain a compact phase space in this way. In particular, it allows to define the global attractor and the bounding graphs as it was done in Section 1.2 .

As we chose the function g to be non-negative, the forcing only 'acts downwards'. We will refer to this case as *'one-sided forcing'*.

The first problem we will encounter is to restrict the number of invariant graphs which can occur. If there are too many, it will be hard to describe a saddle-node bifurcation in detail. Fortunately, there exist general results which allow this, without placing to restrictive conditions on the system. We will discuss these in Section 3.2 (see Theorems 3.2 and 3.3, taken from [**39**] and [**4**]), before giving the proof of Theorem 2.1 . The most convenient of these criteria is to require F to have negative Schwarzian derivative, which ensures that there can be at most three different invariant graphs (Theorem 3.2).[1] However, in the particular situation we consider it will also be sufficient if F is convex on one suitable interval and uniformly contracting on another. More precisely, we will use the following assumption:

(2.5) Suppose that either F is $\mathcal{C}^3$ and has negative Schwarzian derivative, or there exists $c \in (x_-, 0]$, such that $F_{|[-2C,x_-)}$ is uniformly contracting and $F_{|[c,2C)}$ is strictly concave.

Now we can state the following result on the existence of saddle-node bifurcations.

THEOREM 2.1 (Saddle-node bifurcation). *Suppose F and g satisfy (2.2)–(2.5) and let $X = [-2C, 2C]$ and $\beta \in [0, C]$ as above. Then the lower bounding graph of the system (2.1), which we denote by φ^-,[2] is continuous and has negative Lyapunov exponent. Its dependence on β is continuous (in $\mathcal{C}^0$-norm) and monotone: If β is increased then φ^- moves downwards, uniformly on all fibres.*

Further, there exists a critical parameter $\beta_c \in (0, C)$, such that the following holds:

(i) If $\beta < \beta_c$, then there exist exactly two more invariant graphs above φ^-, both of which are continuous. We denote the upper one by φ^+ and the middle one by ψ, such that $\varphi^- < \psi < \varphi^+$. There holds $\lambda(\psi) > 0$ and $\lambda(\varphi^+) < 0$, and the dependence of the graphs on β is continuous and monotone: If β is increased then φ^+ moves downwards, whereas ψ moves upwards, uniformly on all fibres.

(ii) If $\beta = \beta_c$, there exist either one or two more invariant graphs above φ^-. We denote them by ψ and φ^+ again (allowing $\psi = \varphi^+$), where $\psi \leq \varphi^+$. Further, one of the two following holds:

- *ψ equals φ^+ m-a.s. and $\lambda(\psi) = \lambda(\varphi^+) = 0$* (Smooth Bifurcation).[3]
- *$\psi \neq \varphi^+$ m-a.s., $\lambda(\psi) > 0$, $\lambda(\varphi^+) < 0$ and both invariant graphs are non-continuous* (Non-smooth Bifurcation).

In any case, the set $B := [\psi, \varphi^+]$ is compact and the set $\{\theta \in \mathbb{T}^1 \mid \psi(\theta) = \varphi^+(\theta)\}$ is dense in $\mathbb{T}^1$.[4]

(iii) If $\beta > \beta_c$, then φ^- is the only invariant graph.

---

[1] The Schwarzian derivative of a $\mathcal{C}^3$ interval map F is defined as

$$SF := \frac{F'''}{F'} - \frac{3}{2}\left(\frac{F''}{F'}\right)^2 .$$

It is intimately related to the cross ratio distortion of the map (see [**42**]), and this relation is exploited in [**39**] to derive the mentioned statement. This is very similar to the proof of Theorem 3.3 given in Section 3.2 (see Remark 3.5).

[2] We keep the dependence of φ^- on β implicit, same for ψ and φ^+ below.

[3] Of course, the natural possibility here is that ψ and φ^+ are continuous and coincide everywhere. However, there is also a second, rather pathological alternative, which cannot be excluded: It might happen that there exists no continuous invariant graph apart from φ^-, but two semi-continuous invariant graphs ψ and φ^+ which are m-a.s. equal. This is discussed in more detail in Section 3.1 . Whether the bifurcation should really be called smooth in this case is certainly debatable. However, as the non-smooth bifurcations we prove to exist later on all involve non-zero Lyapunov exponents, we prefer this as a working definition in the context of this paper.

[4] A compact set $B \subseteq \mathbb{T}^1 \times X$ is called *pinched*, if for a dense set of θ the set $B_\theta := \{x \in X \mid (\theta, x) \in B\}$ consists of a single point. Thus, the last property could also be stated as 'the set B is pinched'.

The proof of Theorem 2.1, together with some preliminary results which are needed, is given in Section 3.2 .

When F depends on an additional parameter, it is also natural to study the dependence of the critical parameter β_0 on this parameter. We refrain from producing a general statement and just concentrate on the arctan-family (1.1) given in the introduction. Let

$$F_\alpha(x) \ := \ \frac{\arctan(\alpha x)}{\arctan(\alpha)} \ .$$

LEMMA 2.2. *Let $\beta_0(\alpha)$ denote the critical parameter of the saddle-node bifurcation in Theorem 2.1 with $F = F_\alpha$ in (2.1). Then $\alpha \mapsto \beta_0(\alpha)$ is continuous and strictly monotonically increasing in α.*

Again, the proof is given in Section 3.2 . We note that while continuity follows under much more general assumptions, the monotonicity depends on the right scaling of the parameter family, namely on the fact that the fixed points of F_α do not depend on α.

2.2. Sink-source-orbits and the existence of SNA

In this subsection we consider a slightly more general situation than in the last, and suppose that

(2.1) T is a qpf monotone interval map;

(2.2) All fibre maps T_θ are differentiable with derivative DT_θ;

(2.3) $(\theta, x) \mapsto DT_\theta(x)$ is continuous and strictly positive.

In particular, this applies to parameter families which satisfy (2.2)–(2.4).

In order to formulate the statements of this section, we have to introduce different Lyapunov exponents. Let $(\theta, x) \in \mathbb{T}^1 \times X$. Then the *(vertical) finite-time forward* and *backward Lyapunov exponents* are defined as

$$\lambda^+(\theta, x, n) \ := \ \frac{1}{n} \sum_{i=0}^{n-1} \log(DT_{\theta+i\omega}(T^i_\theta(x))) \tag{2.4}$$

and

$$\lambda^-(\theta, x, n) \ := \ -\frac{1}{n} \sum_{i=1}^{n} \log(DT_{\theta-i\omega}(T^{-i}_\theta(x))) \ . \tag{2.5}$$

When dealing with parameter families as in (2.1), we will write $\lambda^\pm(\beta, \theta, x, n)$ for the pointwise finite-time Lyapunov exponents with respect to the map T_β if we want to keep the dependence on the parameter β explicit.

As it is not always possible to ensure that the finite-time exponents converge as $n \to \infty$, we distinguish between upper and lower Lyapunov exponents: The *(vertical) upper forward Lyapunov exponent* of a point $(\theta, x) \in \mathbb{T}^1 \times X$ is defined as

$$\lambda^+(\theta, x) \ := \ \limsup_{n\to\infty} \lambda^+(\theta, x, n) \ . \tag{2.6}$$

Similarly, the *upper backward Lyapunov exponent* is defined as

$$\lambda^-(\theta, x) \ := \ \limsup_{n\to\infty} \lambda^-(\theta, x, n) \ . \tag{2.7}$$

In the same way, we define the *lower forward* and *backward Lyapunov exponents*, replacing lim sup by lim inf:

$$\lambda^+_{\text{low}}(\theta,x) \quad = \quad \liminf_{n\to\infty} \lambda^+(\theta,x,n) \; ; \tag{2.8}$$

$$\lambda^-_{\text{low}}(\theta,x) \quad := \quad \liminf_{n\to\infty} \lambda^-(\theta,x,n) \; . \tag{2.9}$$

Again, we write $\lambda^\pm(\beta,\theta,x), \lambda^\pm_{\text{low}}(\beta,\theta,x)$ if we want to keep the dependence on a parameter β explicit.

For any invariant graph φ, the Birkhoff ergodic theorem implies that for m-a.e. $\theta \in \mathbb{T}^1$ the lim sup and the lim inf coincide (i.e. the respective limits exists and we do not have to distinguish between $\lambda^\pm$ and $\lambda^\pm_{\text{low}}$) and there holds $\lambda^+(\theta,\varphi(\theta)) = -\lambda^-(\theta,\varphi(\theta)) = \lambda(\varphi)$. Further, when φ is continuous the Uniform Ergodic Theorem (e.g. [**41**]) implies that this holds for all $\theta \in \mathbb{T}^1$ and the convergence is uniform on $\mathbb{T}^1$. Now, consider the situation where ψ is an unstable and φ is a stable continuous invariant graph, and there is no other invariant graph in between. Then points on the repeller (or *source*) ψ will have a positive forward and a negative backward Lyapunov exponent, and for points on the attractor (or *sink*) φ it is just the other way around. Further, all points between ψ and φ will converge to φ forwards and to ψ backwards in time, thus moving from source to sink, and consequently both their exponents will be negative. These three cases should be considered as more or less typical. In contrast to this, the remaining possibility of both Lyapunov exponents being positive is rather strange, as it would suggest that the orbit somehow moves from a sink to a source. This motivates the following definition:

DEFINITION 2.3 (Sink-source-orbits). *Suppose T satisfies the assumptions (2.1)–(2.3). Then we call an orbit of T which has both positive forward and backward lower Lyapunov exponent a* ***sink-source-orbit****. If an orbit has both positive forward and backward upper Lyapunov exponent then we call it a* ***weak sink-source-orbit****.*

Obviously, every sink-source-orbit is also a weak sink-source-orbit.

As mentioned in the introduction, the existence of sink-source-orbits is already known for the Harper map (see Footnote 6), where they only occur together with SNA (i.e. in the non-uniformly hyperbolic case, as discussed in Section 1.3.2). This is not a mere coincidence:

THEOREM 2.4. *Suppose T satisfies the assumptions (2.1)–(2.3). Then the existence of a weak sink-source-orbit implies the existence of a SNA (and similarly of a SNR).*

The proof is given in Section 3.3 .

REMARK 2.5. (a) In the proofs of Theorems 2.7 and 2.10 below, we actually construct sink-source-orbits. Thus, for the main purpose of this paper it would not have been necessary to introduce weak sink-source-orbits. However, since the existence of the latter is a much weaker assumption than the existence of a sink-source-orbit (see also (b) and (c) below), it seemed appropriate to state Theorem 2.4 in this way.

(b) In some situations, it is also possible to obtain results in the opposite direction. For example, if M is a minimal set which contains both a SNA and a SNR, then weak sink-source-orbits are dense (even residual) in M. In order to see this, note that, in the above situation, for some constant

$c > 0$ the set M contains a point (θ_1, x_1) with $\lambda^+(\theta_1, x_1) > c$ and a point (θ_2, x_2) with $\lambda^-(\theta_2, x_2) > c$. Due to minimality, it follows that the open sets

$$A_n := \{(\theta, x) \in M \mid \exists m \geq n : \lambda^+(\theta, x, m) > c\}$$

and

$$B_n := \{(\theta, x) \in M \mid \exists m \geq n : \lambda^-(\theta, x, m) > c\}$$

are both dense in M. By Baire's Theorem, their intersection $S := \bigcap_{n \in \mathbb{N}} A_n \cap B_n$ is residual, and obviously every point in S belongs to a weak sink-source-orbit.

(c) The preceding remark becomes false if 'weak sink-source-orbit' is replaced by sink-source-orbit. In fact, it is well-known that SNA may exist in the absence of sink-source-orbits, even if there is a minimal set which contains both an SNA and a SNR. Examples are provided by the Harper map: As we have discussed in Section 1.3.2, the existence of a sink-source-orbit is equivalent to the existence of an exponentially decaying eigenfunction for the corresponding Schrödinger operator. However, there are situations in which, for certain energies in the spectrum of $H_{\lambda,\theta}$ (which does not depend on θ), there exist no such 'localised' eigenfunctions, independent of θ. This follows for example from Theorem 5 in [**46**], together with the concept of Aubry-duality, which is explained in Section 2 of the same paper (the original source is [**47**]). The fact that there is a (unique) minimal set which contains both an SNA and an SNR in these examples is shown in [**1**, Section 4.17].

For a more detailed discussion of the implications of spectral-theoretic results for the Harper map, we also refer to [**43**] (this particular issue is addressed in Section V(C)).

An observation which was made frequently in numerical studies of SNA is a very unusual distribution of the finite-time Lyapunov exponents. The interesting fact is that although in the limit all observed Lyapunov exponents were negative, the distribution of the finite-time Lyapunov exponents still showed a rather large proportion of positive values, even at very large times (see [**11**],[**19**]). Of course, the existence of a sink-source orbit could be a possible explanation for such a behaviour. On the other hand, we can also use information about the finite-time Lyapunov exponents to establish the existence of a sink-source-orbit, and this will play a key role in the proof of our main results:

LEMMA 2.6. *Let I be a compact metric space $\mathbb{R}$ and $(T_\beta)_{\beta \in I}$ be a parameter family of qpf monotone interval maps which all satisfy the assumptions (2.1)–(2.3) above. Further, assume that the dependence of the maps T_β and $(\theta, x) \mapsto DT_{\beta,\theta}(x)$ on β is continuous (w.r.t. the topology of uniform convergence).*

Suppose there exist sequences of integers $l_1^-, l_2^-, \ldots \nearrow \infty$ and $l_1^+, l_2^+, \ldots \nearrow \infty$, a sequence $(\theta_p, x_p)_{p \geq 1}$ of points in $\mathbb{T}^1 \times X$ and a sequence of parameters $(\beta_p)_{p \geq 1}$, such that for all $p \in \mathbb{N}$ there holds

$$\lambda^+(\beta_p, \theta_p, x_p, j) \; > \; c \quad \forall j = 1, \ldots, l_p^+$$

and

$$\lambda^-(\beta_p, \theta_p, x_p, j) \; > \; c \quad \forall j = 1, \ldots, l_p^-$$

for some constant $c > 0$. Then there is at least one $\beta_0 \in I$, such that there exists a sink-source-orbit (and thus a SNA-SNR-pair) for the map T_{β_0}.

PROOF. In fact, the statement is a simple consequence of compactness and continuity: By going over to suitable subsequences if necessary, we can assume that the sequences $(\theta_p)_{p\geq 1}, (x_p)_{p\geq 1}$ and $(\beta_p)_{p\geq 1}$ converge. Denote the limits by θ_0, x_0 and β_0, respectively.

Now, due to the assumptions on T_β and $DT_{\beta,\theta}(x)$ the functions $(\beta, \theta, x) \mapsto \lambda^\pm(\beta, \theta, x, j)$ are continuous for each fixed $j \in \mathbb{N}$. Thus, we obtain

$$\lambda^\pm(\beta_0, \theta_0, x_0, j) \;=\; \lim_{p\to\infty} \lambda^\pm(\beta_p, \theta_p, x_p, j) \;\geq\; c \quad \forall j \in \mathbb{N} ,$$

such that

$$\lambda^\pm_{\text{low}}(\beta_0, \theta_0, x_0) \;=\; \liminf_{j\to\infty} \lambda^\pm(\beta_0, \theta_0, x_0, j) \;\geq\; c \;>\; 0 .$$

Hence, the orbit of (θ_0, x_0) is a sink-source-orbit for the map T_{β_0}. □

2.3. Non-smooth bifurcations

In order to formulate the results concerning the non-smoothness of bifurcations and the existence of SNA, we first have to quantify the qualitative features of the functions F and g which were used in the discussion in Section 1.4 . Some of the assumptions we will make below are quite specific and could in principle be formulated in a more general way. However, as the proofs of Theorems 2.7 and 2.10 are quite involved anyway, we refrain from introducing any more additional parameters, even if this could lead to slightly more flexible results. As we have mentioned before, our main goal here is just to show that the presented approach does lead to rigorous results at all, we do not aim for the greatest possible generality. Hence, we content ourself here to provide a statement which applies, after suitable rescaling and reparametrisation, to at least two of the main examples from the introduction (see Sections 2.4.1 and 2.4.2).

First of all, we will suppose that γ and α are positive constants which satisfy

(2.1) $$\gamma \;\leq\; 1/16 ;$$

(2.2) $$\sqrt{\alpha} \;>\; 4/\gamma \;\geq\; 64 .$$

Further, we will assume (in addition to (2.2)–(2.5)), that

(2.3) $F([-3,3]) \subseteq [-3/2, 3/2]$ (in other words $C = 3/2$ in (2.3));

(2.4) $F(0) = 0$ and $F(\pm x_\alpha) = \pm x_\alpha$ where $x_\alpha \;:=\; 1 + \frac{2}{\sqrt{\alpha}}$;

(2.5) $2\alpha^{-2} \;\leq\; F'(x) \;\leq\; \alpha^2 \qquad \forall x \in [-3,3]$;

(2.6) $F'(x) \;\geq\; 2\alpha^{\frac{1}{2}} \qquad \forall x \in \overline{B_{\frac{2}{\alpha}}(0)}$;

(2.7) $F'(x) \;\leq\; \frac{1}{2}\alpha^{-\frac{1}{2}} \qquad \forall x : |x| \geq \gamma$;

(2.8) $F(\frac{1}{\alpha}) \;\geq\; 1 - \gamma$ and $F(-\frac{1}{\alpha}) \;\leq\; -(1-\gamma)$.

Finally, we will require that

(2.9) $g : \mathbb{T}^1 \to [0,1]$ has the unique maximum $g(0) = 1$

and for some constants $L_1, L_2 > 0$ there holds

$$g \text{ is Lipschitz-continuous with Lipschitz constant } L_1 \ ; \tag{2.10}$$

$$g(\theta) \ \leq \ \max\{1-3\gamma, 1-L_2\cdot d(\theta,0)\} \ , \tag{2.11}$$

where d denotes the usual Euclidean distance on the circle. Essentially, this quantifies the properties which we have already mentioned in Section 1.4: F has three fixed points (2.4), acts highly expanding close to 0 (2.6) and highly contracting further away (2.7). Thus, the expanding region $\mathcal{E}$ from Section 1.4 corresponds to $\mathbb{T}^1 \times \overline{B_{\frac{2}{\alpha}}(0)}$, whereas the contracting region $\mathcal{C}$ corresponds to $\mathbb{T}^1 \times [\gamma, 3]$. Further, (2.8) ensures that $\mathbb{T}^1 \times \overline{B_{\frac{1}{\alpha}}(0)}$ is mapped over itself in a very strong sense, and finally condition (2.11) makes precise what we meant when speaking of a 'sharp peak' before.

The last assumption we need is a Diophantine condition on the rotation number ω. We use the notation

$$\omega_n := n\omega \bmod 1 \tag{2.12}$$

and suppose that there exist constants $c, d > 0$, such that

$$d(\omega_n, 0) \ \geq \ c\cdot n^{-d} \quad \forall n \in \mathbb{N} \ . \tag{2.13}$$

(Here $d(\theta, \theta')$ denotes the usual Euclidean distance of two points $\theta, \theta' \in \mathbb{T}^1$.)

THEOREM 2.7. *Suppose α, γ, F and g are chosen such that (2.2)–(2.5) and (2.1)–(2.11) hold. Further, assume that ω satisfies the Diophantine condition (2.13) and let*

$$T_\beta(\theta, x) \ = \ (\theta+\omega, F(x) - \beta g(\theta))$$

as in (2.1). Let $\beta_c \in (0, 3/2)$ be the critical parameter of the saddle-node bifurcation described in Theorem 2.1 . Then there exist constants $\gamma_0 = \gamma_0(L_1, L_2, c, d) > 0$ and $\alpha_0 = \alpha_0(L_1, L_2, c, d) > 0$ with the following property:

If $\gamma < \gamma_0$ and $\alpha > \alpha_0$, then there exists a sink-source-orbit for the system T_{β_c}. Consequently, there exists a SNA (the invariant graph φ^+ in Theorem 2.1(ii)) and a SNR (ψ in Theorem 2.1(ii)), and both objects have the same essential closure.[5]

The proof of this theorem is given in the Sections 4–6, an outline of the strategy is given at the beginning of Section 4.

REMARK 2.8. (a) We remark that the existence of a sink-source-orbit in the parameter family T_β in the above theorem does not depend on the statement of Theorem 2.1. Even if the assumptions (2.2)–(2.5) are dropped and Theorem 2.1 no longer applies, we still obtain the existence of a parameter β_c for which T_{β_c} has a sink-source-orbit and a SNA-SNR-pair, provided γ is sufficiently small and α sufficiently large. However, in this case β_c is not necessarily unique anymore. Further, it is not possible to say whether it is a bifurcation parameter, nor to control the number of invariant graphs which might occur.

(b) The dependence of γ_0 and α_0 on L_1, L_2, c and d can be made explicit. More precisely, the conditions which have to be satisfied are (5.1), (5.1), (5.2) and (6.1)–(6.4). Conditions (5.1) and (5.2) are somewhat implicit, but once the parameters u and v are fixed according to (6.1)–(6.3), explicit formulas can be derived from the proof of Lemma 5.11 .

[5]See Section 3.1 for the definition of the essential closure.

(c) Numerical observations (as well as the statement of the above theorem) suggest that there might be a critical parameter α^*, such that the saddle-node bifurcation in the family $\tilde{T}_{\alpha,\beta}$ with fixed α is smooth whenever $\alpha < \alpha^*$ and non-smooth whenever $\alpha > \alpha^*$. However, whether this is really the case is completely open.

As we have mentioned in Section 1.4, the sharp peak of the forcing function leads to a localisation of the sink-source-orbit. In fact, its construction in the later sections yields enough information to determine it precisely:

ADDENDUM 2.9. *In the situation of Theorem 2.7 denote the SNA by φ^+ and the SNR by ψ. Then the point $(\omega, \varphi^+(\omega))$ belongs to a sink-source-orbit.*[6] *Further, this sink-source-orbit is contained in the intersection $\Phi^+ \cap \Psi$ (which means $\varphi^+(\omega) = \psi(\omega)$).*

The proof is given in Section 6.

Next, we turn to the existence of SNA in symmetric systems. In order to do so, we have to modify the assumptions on F and g. First of all, instead of (2.2) and (2.5) we will assume that

(2.14) $g : \mathbb{T}^1 \to [-1, 1]$ is continuous and has the unique maximum $g(0) = 1$;

(2.15) F is $\mathcal{C}^3$ and has negative Schwarzian derivative.

(The alternative in (2.5) only works for one-sided forcing). Further, we will require the following symmetry conditions

$$F(-x) = -F(x) \; ; \tag{2.16}$$

$$g(\theta + \tfrac{1}{2}) = -g(\theta) \; . \tag{2.17}$$

Finally, (2.11) will be replaced by

$$|g(\theta)| \leq \max\{1 - 3\gamma, 1 - L_2 \cdot d(\theta, \{0, \tfrac{1}{2}\})\} \; . \tag{2.18}$$

Note that (2.16) and (2.17) together imply that the map $T = T_\beta$ given by (2.1) has the following symmetry property:

$$-T_\theta(x) = T_{\theta + \frac{1}{2}}(-x) \tag{2.19}$$

Now suppose that φ is a T-invariant graph. Then due to (2.19) the graph given by

$$\overline{\varphi}(\theta) := -\varphi(\theta + \tfrac{1}{2}) \tag{2.20}$$

is invariant as well. In particular, this implies that the upper and lower bounding graphs satisfy $\varphi^+(\theta) = -\varphi^-(\theta + \frac{1}{2})$, and if one of these graphs undergoes a bifurcation, then the same must be true for the second one as well. As the negative Schwarzian derivative of F will allow us to conclude that there is only one other invariant graph ψ apart from the bounding graphs $\varphi^\pm$, this implies that any possible collision between invariant graphs has to involve all three invariant graphs at the same time and must therefore be a pitchfork bifurcation. However, as we have mentioned before, due to the lack of monotonicity in the symmetric setting we cannot ensure that there is a unique bifurcation point. Nevertheless, we obtain the following result concerning the existence of SNA with symmetry:

[6] There might be more than one sink-source-orbit, but this is the particular one which we will construct in the later sections.

THEOREM 2.10. *Suppose γ, α, F and g are chosen, such that (2.1)–(2.8), (2.10) and (2.14)–(2.18) hold. Further, assume ω satisfies the Diophantine condition (2.13) and let*

$$T_\beta(\theta, x) \; = \; (\theta + \omega, F(x) - \beta g(\theta))$$

as in (2.1). Then there exist constants $\gamma_0 = \gamma_0(L_1, L_2, c, d) > 0$ and $\alpha_0 = \alpha_0(L_1, L_2, c, d) > 0$ with the following property:

If $\gamma < \gamma_0$ and $\alpha > \alpha_0$, then there is a parameter β_c such that there exist two SNA φ^- and φ^+ and a SNR ψ with $\varphi^- \leq \psi \leq \varphi^+$ for T_{β_c}. Further, there holds $\overline{\Phi^-}^{ess} = \overline{\Psi}^{ess} = \overline{\Phi^+}^{ess}$, and the invariant graphs satisfy the symmetry equations

$$\varphi^-(\theta) = -\varphi^+(\theta + \tfrac{1}{2}) \quad \text{and} \quad \psi(\theta) = -\psi(\theta + \tfrac{1}{2}) \; .$$

REMARK 2.11. As in Theorem 2.7, the dependence of γ_0 and α_0 on L_1, L_2, c and d can be made explicit (compare Remark 2.8(b)). The conditions which have to be satisfied are (5.1), (5.1), (5.2), (6.1)–(6.4), (7.1) and (7.2).

2.4. Application to the parameter families

The assumptions on F and g used in Theorems 2.7 and 2.10 are somewhat technical and might seem very restrictive. However, in this subsection we will see that they are more flexible that they might look like on first sight (although there are surely some constraints). In particular, after performing some surgery we can apply them at least to two of the parameter families from the introduction, namely the arctan-family with additive forcing and the Harper map. In both cases, the respective parameters have to chosen sufficiently large, but of course this goes perfectly well with the statement of Theorem 2.7 . As a consequence, the respective corollaries become a lot easier to formulate.

The qpf Arnold map then demonstrates the limits of Theorem 2.7, since it is not possible to apply the result in this case. This is briefly discussed in Subsection 2.4.3 .

2.4.1. Application to the arctan-family. Applied to the arctan-family with additive forcing, Theorem 2.7 yields the following:

COROLLARY 2.12. *Suppose ω satisfies the Diophantine condition (2.13). Then there exists $\alpha_0 = \alpha_0(c, d)$ such that for all $\alpha > \alpha_0$ the system $T_{\alpha,\beta}$ given by (1.1) undergoes a non-smooth saddle-node bifurcation as the parameter β is increased from 0 to 1.*

REMARK 2.13. As already mentioned in Section 1.3.1, the above statement remains true if the arctan in (1.1) is replaced by the map $x \mapsto \frac{x}{1+|x|}$, or any other function which has similar scaling properties. This will become obvious in the following proof, but we refrain from producing a more general statement here.

Proof of Corollary 2.12 . Since the system (1.1) does not satisfy (2.4), we cannot apply Theorem 2.7 directly. Therefore, we start by considering a slightly rescaled version of (1.1). Let

$$\tilde{F}_\alpha(x) \; := \; C(\alpha) \cdot \arctan(\alpha^{\frac{4}{3}} x) \quad \text{where} \quad C(\alpha) \; := \; \frac{1 + \frac{2}{\sqrt{\alpha}}}{\arctan(\alpha^{\frac{4}{3}} + 2\alpha^{\frac{5}{6}})}$$

and

$$g(\theta) \; := \; 1 - \sin(\pi\theta) \; .$$

Note that $\tilde{F}_\alpha$ always satisfies (2.4). The important thing we have to ensure is that whenever we fix a suitably small γ, such that (2.1), (2.11) and any additional smallness conditions on γ which appear later on are satisfied, then (2.8) holds for all sufficiently large values of α. This means that we can first fix γ, and then ensure that all inequalities involving α alone or both α and γ, such as (2.2), hold by choosing α sufficiently large, without worrying about (2.8). However, in this particular case it is easy to see that $\tilde{F}_\alpha(\frac{1}{\alpha}) = C(\alpha) \cdot \arctan(\alpha^{\frac{1}{3}}) \to 1$ as $\alpha \to \infty$ (note that $\lim_{\alpha\to\infty} C(\alpha) = \frac{2}{\pi}$), which is exactly what we need.

Now, if γ is chosen small enough $g(\theta) = |1 - \sin(\pi\theta)|$ clearly satisfies (2.11), for example with $L_2 := 2$. The Lipschitz-constant L_1 is π. Thus, it remains to check the assumptions on the derivative of $\tilde{F}_\alpha$. To that end, note that

$$\tilde{F}'_\alpha(x) \;=\; C(\alpha) \cdot \frac{\alpha^{\frac{4}{3}}}{1+\alpha^{\frac{8}{3}}x^2}$$

We have $\tilde{F}'_\alpha(0) \sim \alpha^{\frac{4}{3}}$, $\tilde{F}'_\alpha(\frac{2}{\alpha}) \sim \alpha^{\frac{2}{3}}$ and $\tilde{F}'_\alpha(\gamma) \sim \alpha^{-\frac{4}{3}}$ for each fixed $\gamma > 0$ as $\alpha \to \infty$. Therefore, the conditions (2.5),(2.6) and (2.7) will always be satisfied when α is large enough. Consequently we can apply Theorem 2.7 and obtain that there exists some $\tilde{\alpha}_0$ such that for all $\alpha \geq \tilde{\alpha}_0$ the parameter family

$$\tilde{T}_{\alpha,\beta} \;:\; (\theta, x) \mapsto (\theta + \omega, \tilde{F}_\alpha(x) - \beta g(\theta))$$

undergoes a non-smooth pitchfork bifurcation (in the sense of Theorem 2.1) as β is increased from 0 to 3/2.

Now denote the map given by (1.1) by $T_{\alpha,\beta}$. We claim that there exists a monotonically increasing function $\sigma : \mathbb{R}^+ \to \mathbb{R}^+$ and a function $\tau : \mathbb{R}^+ \to \mathbb{R}^+$ such that $T_{\alpha,\beta}$ is smoothly conjugate to $\tilde{T}_{\sigma(\alpha),\tau(\alpha)\beta}$. Consequently, the parameter family $T_{\alpha,\beta}$ equally exhibits non-smooth saddle-node bifurcations if α is chosen sufficiently large (larger than $\sigma^{-1}(\tilde{\alpha}_0)$).

In order to define σ, it is convenient to introduce an intermediate parameter family $\widehat{T}_{\alpha,\beta}$ with fibre maps

$$\widehat{T}_{\alpha,\beta,\theta}(x) \;=\; \arctan(\alpha x) - \beta g(\theta)\ .$$

We let $h_1(\theta, x) = (\theta, \arctan(\alpha)x)$, $\sigma_1(\alpha) = \arctan(\alpha)^{-1}\alpha$ and $\tau_1(\alpha) = \arctan(\alpha)$. Then

$$T_{\alpha,\beta} \;=\; h_1^{-1} \circ \widehat{T}_{\sigma_1(\alpha),\tau_1(\alpha)\beta} \circ h_1\ ,$$

such that $T_{\alpha,\beta} \sim \widehat{T}_{\sigma_1(\alpha),\tau_1(\alpha)\beta}$, where $\sim$ denotes the existence of a smooth conjugacy.

On the other hand, let $h_2(\theta, x) = (\theta, C(\alpha)^{-1}x)$, $\sigma_2(\alpha) = C(\alpha)\alpha^{\frac{4}{3}}$ and $\tau_2(\alpha) = C(\alpha)^{-1}$. Again, a simple computation yields

$$\tilde{T}_{\alpha,\beta} \;=\; h_2^{-1} \circ \widehat{T}_{\sigma_2(\alpha),\tau_2(\alpha)\beta} \circ h_2\ .$$

As σ_1 and σ_2 are both strictly monotonically increasing and therefore invertible, this implies $\widehat{T}_{\alpha,\beta} \sim \tilde{T}_{\sigma_2^{-1}(\alpha),\tau_2(\sigma_2^{-1}(\alpha))^{-1}\beta}$ and consequently

$$T_{\alpha,\beta} \;\sim\; \widehat{T}_{\sigma_1(\alpha),\tau_1(\alpha)\beta} \;\sim\; \tilde{T}_{\sigma_2^{-1}\circ\sigma_1(\alpha),\tau_2(\sigma_2^{-1}\circ\sigma_1(\alpha))^{-1}\tau_1(\alpha)\beta}\ .$$

Hence, we can define $\sigma = \sigma_2^{-1} \circ \sigma^1$ and $\tau = \frac{\tau_1}{\tau_2\circ\sigma_2^{-1}\circ\sigma_1}$ as claimed.

Finally, since F_α has the fixed point $x_+ = 1$, $g(0) = 1$ and we are in the case of one-sided forcing, it can easily be seen that the bifurcation must take place before $\beta = 1$ (meaning that the critical parameter β_c given Theorem 2.1 is strictly smaller than one). For larger β-values all orbits eventually end up below the 0-line and consequently converge to the lower bounding graph φ^-, such that this is the only invariant graph. (Compare with the proof of Theorem 2.1 in Section 3.2.) This completes the proof.

□

2.4.2. Application to the Harper map. We want to emphasise that we do not claim any originality for the presented results on the Harper map. Our aim is merely to demonstrate the flexibility of our general statements by applying them to this well-known family. For the particular case of the Harper map there surely exist more direct and elegant ways to produce such results, starting with [**1**]. Usually such results require more regularity than we use here (the potentials we can treat are only Lipschitz-continuous), but from the physics point of view this is surely the more interesting case anyway. Further, although potentials which are only Lipschitz-continuous are not explicitly treated in [**33**], the methods developed there surely allow to do this as well. Thus, the real achievement here is rather to show that the underlying mechanism for non-smooth bifurcations is in principle the same in the Harper map as in other parameter families, like the arctan-family with additive forcing, despite the very particular structures which distinguish Schrödinger cocycles from other models.

As Theorem 2.7 is tailored-made for qpf interval maps, its application to the Harper map is somewhat indirect. This means that we have to perform a number of modifications before the system in (1.6) is in a form which meets the assumptions of the theorem. First of all, we remark that the dynamics of (1.6) are equivalent to those of the map

$$(\theta, x) \mapsto \left(\theta + \omega, \frac{-1}{x} + E - \lambda V(\theta + \omega)\right) \tag{2.1}$$

defined on $\mathbb{T}^1 \times \overline{\mathbb{R}}$. In order to see this, note that, by taking the inverse and replacing ω by $-\omega$ in (2.1), we obtain the system

$$(\theta, x) \mapsto \left(\theta + \omega, \frac{-1}{x - E + \lambda V(\theta)}\right) \tag{2.2}$$

Using the change of variables $x \mapsto \tan(-x)$, this yields (1.6). The proof of Corollary 2.14 below will mainly consist in showing that there exists a parameter family of qpf interval maps which satisfies the assumptions of Theorem 2.7, such that it exhibits a non-smooth saddle-node bifurcation, and which is at the same time conjugated to (2.1), provided that both systems are restricted to the relevant part of the phase space in which the bifurcation takes place.

In order to proceed, we will now first take the rather atypical viewpoint of fixing E and considering λ as the bifurcation parameter (whereas usually in the study of Schrödinger cocycles the coupling constant λ is fixed and the spectral parameter E is varied). However, in this particular situation the two viewpoints are actually equivalent and the analogous result from the standard viewpoint can be recovered afterwards. More precisely, we first show that Theorem 2.7 implies the following:

COROLLARY 2.14. *Suppose ω satisfies (2.13) and the potential V is non-negative, Lipschitz-continuous and decays linearly in a neighbourhood of its unique maximum. Then there exists a constant $E_0 = E_0(V, c, d)$ with the following property:*

If $E \geq E_0$, then there exists a unique parameter $\lambda_c = \lambda_c(E)$, such that for all $\lambda \in [0, \lambda_c]$ there exist exactly two invariant graphs for the system (1.6) (and likewise for (2.1)), one with positive and one with negative Lyapunov exponent. If $\lambda < \lambda_c$ then both these graphs are continuous, if $\lambda = \lambda_c$ they are non-continuous (i.e. a SNA and a SNR) and have the same topological closure. Furthermore, the mapping $E \mapsto \lambda_c(E)$ is strictly monotonically increasing.

Due to the monotone dependence of $\lambda_c(E)$ on E, Corollary 2.14 immediately implies

COROLLARY 2.15. *Suppose ω, V and E_0 are chosen as in Corollary 2.14 and let $\lambda_0 := \lambda_c(E_0)$. Then the following holds:*

If $\lambda \geq \lambda_0$, then there exists a unique parameter $E_c = E_c(\lambda) \geq E_0$, such that for all $E \geq E_c$ there exist exactly two invariant graphs for the system (1.6) (and likewise for (2.1)), one with positive and one with negative Lyapunov exponent. If $E > E_c$ then both these graphs are continuous, if $E = E_c$ they are non-continuous (i.e. a SNA and a SNR) and have the same topological closure. The mapping $\lambda \mapsto E_c(\lambda)$ is the inverse of the mapping $E \mapsto \lambda_c(E)$.

We remark that the Harper map can be viewed as a qpf circle homeomorphism (by identifying $\mathbb{R} \cup \{\infty\}$ with $\mathbb{T}^1$). Since we do not want to introduce rotation numbers for such systems here, we do not speak more precisely about what happens if E is decreased beyond E_c (or λ is increased beyond λ_c) and just mention that in this case the rotation number starts to increase and becomes non-zero for $E < E_c$. Invariant graphs and even continuous invariant curves may exist in this situation, but they will have a different homotopy type (i.e. they 'wind around the torus' in the vertical direction).

Proof of Corollary 2.14. In the following we always assume that the parameter E is chosen sufficiently large, without further mentioning. (In particular, most of the statements below are only true for large E.)

As the two systems (1.6) and (2.1) are equivalent (as mentioned above), it suffices to show that the statement is true for (2.1). Further, for the sake of simplicity we assume that V is normalised, i.e. $\sup_{\theta \in \mathbb{T}^1} V(\theta) = 1$. Let $\alpha := E^{3/2}$ and

$$F_1(x) := -1/x + E .$$

Then

$$F_1([1/E, 2/E]) \supseteq [1/E, 2/E] =: I_1$$

and

$$F_1([3E/4, E]) \subseteq [3E/4, E] =: I_2 .$$

Further F_1 is uniformly expanding on I_1 and uniformly contracting on I_2. As F_1 is strictly concave on $(0, \infty)$, it follows that $F_{1|(0,\infty)}$ has exactly two fixed points $x_1 \in I_1$ and $x_2 \in I_2$.

Let $s := \frac{1+2/\sqrt{\alpha}}{x_2 - x_1}$ and $h(x) := (x - x_1) \cdot s$. Note that we have

$$s \in [1/E, 2/E] .$$

As h sends x_1 to 0 and x_2 to $1 + 2/\sqrt{\alpha}$, the map

$$F_2(x) := h \circ F_1 \circ h^{-1}(x) = \frac{-s}{x/s + x_1} + s \cdot (E - x_1) \tag{2.3}$$

has fixed points 0 and $1+2/\sqrt{\alpha}$. In addition, if $\gamma \in (0,1)$ is fixed, then it is easy to check that on the one hand

$$F_2'(x) \in [1/4E^2, 4/\gamma^2E^2] \subseteq \left[\alpha^{-2}, \frac{1}{2}\alpha^{-\frac{1}{2}}\right] \qquad \forall x \in [\gamma, 1+2/\sqrt{\alpha}]$$

and on the other hand

$$F_2'(x) \in [E/16, E^2] \subseteq [2\alpha^{\frac{1}{2}}, \alpha^2] \qquad \forall x \in [0, 2/\alpha] \ .$$

(Always assuming that E is sufficiently large.) Further, there holds

$$F_2(1/\alpha) = \frac{-s}{1/\alpha s + x_1} + s\cdot(E - x_1) \geq -4/\sqrt{E} + 1 - 4/E^2 \stackrel{E\to\infty}{\longrightarrow} 1 \ ,$$

such that we can assume $F_2(1/\alpha) \geq 1-\gamma$.

Due to the definition of F_2 in (2.3) and as h is affine with slope s, the map $H : (\theta, x) \mapsto (\theta, h(x))$ smoothly conjugates (2.1) with

$$(\theta, x) \mapsto (\theta+\omega, F_2(x) - s\lambda V(\theta+\omega)) \ . \tag{2.4}$$

Now we choose a $\mathcal{C}^1$-map $F : [-3,3] \to [-\frac{3}{2}, \frac{3}{2}]$, such that $F_{|[0,1+2/\sqrt{\alpha}]} = F_{2|[0,1+2/\sqrt{\alpha}]}$ and which satisfies the requirements (2.4)–(2.8). This is possible, since we have shown above that $F_{2|[0,1+2/\sqrt{\alpha}]}$ has all the required properties. In addition, F can be chosen such that it is strictly concave on $(0,3]$, has a unique fixed point x_- in $[-3,0)$ and is uniformly contracting on $[-3, x_-]$. Consequently, it satisfies the second alternative of (2.5). Further, if we let $g(\theta) := V(\theta+\omega)$ and $\beta := s\lambda$ and define

$$T_\beta(\theta, x) = (\theta+\omega, F(x) - \beta g(\theta))$$

as in Theorem 2.7, then H conjugates T_β restricted to $\mathbb{T}^1 \times [0, 1+2/\sqrt{\alpha}]$ and $S_{E,\lambda}$ restricted to $\mathbb{T}^1 \times [x_1, x_2]$, where $S_{E,\lambda}$ denotes the map given by (2.1).

The parameter family T_β satisfies all the requirements of Theorem 2.7. Hence, we obtain the existence of a critical parameter β_c, which is the bifurcation parameter in a non-smooth saddle-node bifurcation. Further, due to the monotonicity described in Theorem 2.1(i), the T_β-invariant graphs ψ and φ^+ are always contained in $\mathbb{T}^1 \times [0, 1+2/\sqrt{\alpha}]$. Consequently their preimages under H, which we denote by $\widehat{\psi}$ and $\widehat{\varphi}^+$, are $S_{E,\lambda}$-invariant and contained in $\mathbb{T}^1 \times (0, +\infty)$. Therefore, the parameter family $S_{E,\lambda}$ equally undergoes a non-smooth saddle-node bifurcation with critical parameter $\lambda_c = \beta_c/s$.

In order to complete the proof only two things remain to be shown: The monotonicity of $E \mapsto \lambda_c(E)$ and the fact that for all $\lambda \leq \lambda_c$ the two graphs $\widehat{\psi}$ and $\widehat{\varphi}^+$ are indeed the only ones for the system $S_{E,\lambda}$. In order to see the latter, we note that restricted to $[0,+\infty)$ all the fibre maps of $S_{E,\lambda}$ are strictly concave, such that there can be only two invariant graphs in $\mathbb{T}^1 \times [0, +\infty)$. However, as $S_{E,\lambda}$ maps $\mathbb{T}^1 \times [-\infty, 0)$ into the forward invariant set $[\varphi^+, +\infty)$, there cannot be any other invariant graphs in $\mathbb{T}^1 \times [-\infty, 0)$ either. (Of course, the same conclusion also follows by considering the associated $\mathrm{SL}(2,\mathbb{R})$-cocycle: Due to the non-zero Lyapunov exponent there exists an invariant splitting into stable and unstable subspaces. These correspond exactly to the two invariant graphs above and there will be no others (compare Section 1.3.2).)

In order to see the strict monotonicity of $E \mapsto \lambda_c(E)$, fix $\epsilon > 0$ and suppose that $E_2 = E_1 + 3\epsilon$ and $\lambda' < \lambda_c(E_1) + \epsilon$. Then $S_{E_1, \lambda_c(E_1)-\epsilon}$ has a continuous invariant graph φ_1^+ contained in $\mathbb{T}^1 \times (0, +\infty)$, and some iterate of $S_{E_1,\lambda_c(E_1)-\epsilon}$

acts uniformly contracting in the vertical direction on $[\varphi_1^+, \infty)$. (This follows from the Uniform Ergodic Theorem in combination with the fact that $DS_{E,\lambda,\theta}(x) = 1/x^2$ is decreasing in x.) However, since

$$S_{E_2,\lambda',\theta}(x) \geq S_{E_1,\lambda_c(E_1)-\epsilon,\theta}(x) \quad \forall x \in [\varphi_1^+, +\infty) ,$$

(recall that we assumed V to be normalised) this implies that $S_{E_2,\lambda'}$ maps $[\varphi_1^+, \infty)$ into itself and the respective iterate of $S_{E_2,\lambda'}$ also acts uniformly contracting in the vertical direction on this set. (Note that $DS_{E,\lambda,\theta}$ does not depend on the parameters E and λ.) Consequently $S_{E_2,\lambda'}$ has an attracting and continuous invariant graph contained in $[\varphi_1^+, +\infty)$. As this is true for all $\lambda' \leq \lambda_c(E_1)+\epsilon$, this implies $\lambda_c(E_2) > \lambda_c(E_1) + \epsilon$.

□

2.4.3. Remarks on the qpf Arnold circle map and Pinched systems. We neither apply the results from Section 2.3 to Pinched skew products nor the the qpf Arnold map, but for very different reasons. In the case of Pinched skew products, this would seem like using a sledgehammer to crack a nut. In these systems the existence of SNA can be established by a few short and elegant arguments, making use of their particular structure (see [**2**] and [**4**]). Even the exponential evolution of peaks can be described in a much more simple way in this setting, a fact which was used in [**35**] to study the topological structure of SNA in Pinched skew products. (In fact, this preceding result and the striking similarities between the pictures in Figures 1.2 and 1.6, which strongly suggested some common underlying pattern, were the starting point for the work presented here). In principle it is possible to view the SNA in these systems as being created in non-smooth bifurcations, as this is done in [**48**]. However, as treating them with the methods presented here would even need some additional modifications, we refrain from doing so.

For the case of the qpf Arnold circle map, the situation is completely different. Here it is just not possible to apply our results. The reason for this is the fact that no matter how the parameter $\alpha \in [0, 1]$ in (1.7) is chosen, the maximal expansion rate is always at most two. Further, for any interval of fixed length the uniform contraction rate also remains bounded. Although the derivative goes to zero at $\theta = \frac{1}{2}$ if α is close to 1, a strong contraction only takes place locally. This means that the expansion and contraction rates one can work with will always be moderate and cannot be chosen arbitrarily large by adjusting the parameters. However, this is exactly what would be necessary for the application of Theorem 2.7 . In the case of the forcing function $\theta \mapsto \sin(2\pi\theta)$ used in (1.7), there is also not much hope that a refinement of our methods would yield results. As the simulations in Figure 1.4 indicate, the exponential evolution of peaks is only present in a very weak form in this case. Therefore, it should be doubted that this process can be described in a rigorous way with approximative methods as the ones we use in the proof of Theorem 2.7, which necessarily involve a lot of rough estimates.

However, as already indicated in Section 1.3.3, this might become different if one chooses a more suitable forcing function, and considers for example the parameter family

$$(\theta, x) \mapsto \left(\theta + \omega, x + \tau + \frac{\alpha}{2\pi}\sin(2\pi x) - \beta \cdot \max\{0, 1 - \sigma \cdot d(\theta, 0)\}\right) \tag{2.5}$$

with sufficiently large parameter σ. In this case the exponential evolution of peaks is very distinct again, as one can see in Figure 1.5. Consequently, it should also be possible to treat this situation rigorously. Nevertheless, Theorem 2.7 is not sufficient for this purpose. Changing the forcing function does not have any influence on the expansion and contraction rates, such that these will still be too weak to meet our assumptions. Yet, there is an additional fact which we do not make use of in the proof of Theorem 2.7: In the situation of (2.5) with large σ, the forcing function vanishes almost everywhere, apart from a small neighbourhood of 0. This means that after every visit in this neighbourhood, the expansion, respectively contraction, has a long time to work, without any quasiperiodic influence, before the next return. It seems reasonable to expect that this could be used to make up for the weak expansion and contraction rates, for example by regarding a renormalisation of the original system after a sufficiently large finite time. However, the implementation of this idea is left for the future ... .

2.4.4. SNA's with symmetry. Similar to the proof of Corollary 2.12, it is possible to show that for sufficiently large parameters α the parameter family (1.9) satisfies the assumptions of Theorem 2.10 . This leads to the following

COROLLARY 2.16. *Suppose ω satisfies the Diophantine condition (2.13). Then there exists $\alpha_0 = \alpha_0(c, d)$ such that for all $\alpha > \alpha_0$ there is a parameter $\beta_c = \beta_c(\alpha)$ such that the system (1.9) with parameters α and β_c has two SNA and one SNR, with the properties described in Theorem 2.10, and no other invariant graphs.*

As the details are more or less the same as in Section 2.4.1, we omit the proof.

To the knowledge of the author, this is the first situation where existence of such a triple of intermingled invariant graphs can be described rigorously. Similarly, it is the first example of a qpf monotone interval map without continuous invariant graphs.

CHAPTER 3

Saddle-node bifurcations and sink-source-orbits

The aim of this section is threefold: First, it is to introduce a general setting where a (not necessarily non-smooth) saddle-node bifurcation occurs and can be described rigorously. Secondly, we will show that the presence of a *'sink-source-orbit'* implies the non-smoothness of the bifurcation, and how the existence of such an orbit can be established by approximation with finite trajectories. The construction of such trajectories with the required properties will then be carried out in the succeeding Sections 4 to 6. Finally, before we can start we have to address a subtle issue concerning the definition of invariant graphs:

3.1. Equivalence classes of invariant graphs and the essential closure

The problem we want to discuss is the following: Any invariant graph φ can be modified on a set of measure zero to yield another invariant graph $\tilde{\varphi}$, equal to φ m-a.s. (where m denotes the Lebesgue measure on $\mathbb{T}^1$). We usually do not want to distinguish between such graphs. On the other hand, especially when topology is concerned we sometimes need objects which are well-defined everywhere. So far, this has not been a problem. The bounding graphs of invariant sets defined by (1.4) are well-defined everywhere, and for the definition of the associated measure (1.5) it does not matter. But in general, some care has to be taken. We will therefore use the following convention:

We will consider two invariant graphs as equivalent if they are m-a.s. equal and implicitly speak about equivalence classes of invariant graphs (just as functions in $\mathcal{L}^1_{\text{Leb}}(\mathbb{R})$ are identified if they are Lebesgue-a.s. equal). Whenever any further assumptions about invariant graphs such as continuity, semi-continuity or inequalities between invariant graphs are made, we will understand it in the way that there is at least one representative in each of the respective equivalence classes such that the assumptions are met. All conclusions which are then drawn from the assumed properties will be true for all such representatives.

There is one case where this terminology might cause confusion: It is possible that an equivalence class contains both an upper and a lower semi-continuous graph, but no continuous graph.[1] This rather degenerate case cannot occur when the Lyapunov exponent of the invariant graph is negative (see [**37**], Proposition 4.1), but when the exponent is zero it must be taken into account. To avoid ambiguities, we will explicitly mention this case whenever it can occur.

In order to assign a well defined point set to an equivalence class of invariant graphs, we introduce the *essential closure*:

[1]To get an idea of what could happen, consider the function $f : x \mapsto \sin \frac{1}{x}\ \forall x \neq 0$. By choosing $f(0) = 1$ we can extend it to an upper semi-continuous function, by choosing $f(0) = -1$ to a lower semi-continuous function, but there is no continuous function in the equivalence class.

DEFINITION 3.1. *Let T be a qpf monotone interval map. If φ is an invariant graph, we define its essential closure as*

$$\overline{\Phi}^{ess} := \{(\theta, x) : \mu_\varphi(U) > 0 \ \forall \text{open neighbourhoods } U \text{ of } (\theta, x)\} \,, \tag{3.1}$$

where the associated measure μ_φ is given by (1.5).

Several facts follow immediately from this definition:

- $\overline{\Phi}^{ess}$ is a compact set.
- $\overline{\Phi}^{ess}$ is equal to the topological support $\mathrm{supp}(\mu_\varphi)$ of the measure μ_φ, which in turn implies $\mu_\varphi(\overline{\Phi}^{ess}) = 1$ (see e.g. [**41**]).
- Invariant graphs from the same equivalence class have the same essential closure (as they have the same associated measure).
- $\overline{\Phi}^{ess}$ is contained in every other compact set which contains μ_φ-a.e. point of Φ, in particular in $\overline{\Phi}$.
- $\overline{\Phi}^{ess}$ is forward invariant under T.[2]

3.2. Saddle-node bifurcations: Proof of Theorem 2.1

As mentioned, the first problem we have to deal with is to restrict the number of invariant graphs which can occur. If there are too many, it will be hard to describe a saddle-node bifurcation in detail. However, there is a result which is very convenient in this situation:

THEOREM 3.2 (Theorem 4.2 in [**39**]). *Suppose T is a qpf monotone interval map and all fibre maps T_θ are $\mathcal{C}^3$. Further assume $(\theta, x) \mapsto DT_\theta(x)$ is continuous and all fibre maps have strictly positive derivative and strictly negative Schwarzian derivative (see Footnote 1). Then there are three possible cases:*

(i) There exists one invariant graph φ with $\lambda(\varphi) \leq 0$.
(ii) There exist two invariant graphs φ and ψ with $\lambda(\varphi) < 0$ and $\lambda(\psi) = 0$.
(iii) There exist three invariant graphs $\varphi^- \leq \psi \leq \varphi^+$ with $\lambda(\varphi^-) < 0$, $\lambda(\psi) > 0$ and $\lambda(\varphi^+) < 0$.

Regarding the topology of the invariant graphs, there are the following possibilities:

(i)' If the single invariant graph has negative Lyapunov exponent, it is always continuous. Otherwise the equivalence class contains at least an upper and a lower semi-continuous representative.
(ii)' The upper invariant graph is upper semi-continuous, the lower invariant graph lower semi-continuous. If φ is not continuous and ψ (as an equivalence class) is only semi-continuous in one direction, then $\overline{\Phi}^{ess} = \overline{\Psi}^{ess}$.
(iii)' ψ is continuous if and only if φ^+ and φ^- are continuous. Otherwise φ^- is at least lower semi-continuous and φ^+ is at least upper semi-continuous. If ψ not lower semi-continuous then $\overline{\Phi^-}^{ess} = \overline{\Psi}^{ess}$, if ψ is not upper semi-continuous then $\overline{\Psi}^{ess} = \overline{\Phi^+}^{ess}$.

---

[2]This can be seen as follows: Suppose $x \in \overline{\Phi}^{ess}$ and U is an open neighbourhood of $T(x)$. Then $T^{-1}(U)$ is an open neighbourhood of x, and therefore $\mu_\varphi(U) = \mu_\varphi \circ T^{-1}(U) > 0$. This means $T(x) \in \overline{\Phi}^{ess}$, and as $x \in \overline{\Phi}^{ess}$ was arbitrary we can conclude that $T(\overline{\Phi}^{ess}) \subseteq \overline{\Phi}^{ess}$. On the other hand $T(\overline{\Phi}^{ess})$ is a compact set which contains μ_φ-a.e. point in Φ, therefore $\overline{\Phi}^{ess} \subseteq T(\overline{\Phi}^{ess})$.

Finally, as long as $\lambda(\varphi^-) < 0$ *the graph* ψ *can be defined by*

$$\psi(\theta) := \sup\{x \in X \mid \lim_{n\to\infty} |T^n_\theta(x) - \varphi^-(\theta + n\omega)| = 0\} \; . \tag{3.1}$$

In order to use the alternative assumption in (2.5), we need a similar result for concave fibre maps, which is due to Keller. The main idea of the argument is contained in [**4**]. However, as the statement was never published in this form, we include a proof.

THEOREM 3.3 (G. Keller). *Suppose* T *is a qpf monotone interval map, all fibre maps* T_θ *are differentiable and* $(\theta, x) \mapsto DT_\theta(x)$ *is continuous. Further, assume that there exist measurable functions* $\gamma^\pm : \mathbb{T}^1 \to X$*, such that for all* $\theta \in \mathbb{T}^1$ *the fibre maps* T_θ *are strictly concave on* $I(\theta) = [\gamma^-(\theta), \gamma^+(\theta)] \subseteq X$*. Then there exist at most two invariant graphs taking their values in* $I(\theta)$*, i.e. satisfying*

$$\varphi(\theta) \in I(\theta) \quad \forall \theta \in \mathbb{T}^1 \; . \tag{3.2}$$

If there exist two invariant graphs $\varphi_1 \leq \varphi_2$ *which both satisfy (3.2), then* $\lambda(\varphi_1) > 0$ *and* $\lambda(\varphi_2) < 0$*.*

Further, if the graphs $\gamma^\pm$ *are continuous and are mapped below themselves, meaning that there holds*

$$T_\theta(\gamma^\pm(\theta)) \leq \gamma^\pm(\theta) \quad \forall \theta \in \mathbb{T}^1 \; , \tag{3.3}$$

then either φ_1, φ_2 *are both continuous, or* φ_1 *is lower semi-continuous,* φ_2 *is upper semi-continuous and* $\overline{\Phi_1}^{ess} = \overline{\Phi_2}^{ess}$*. (If there is only one invariant graph which satisfies (3.2), then it always contains an upper and a lower semi-continuous representative in its equivalence class.)*

PROOF. Suppose for a contradiction that there exist three different invariant graphs $\varphi_1 \leq \varphi_2 \leq \varphi_3$ which all satisfy (3.2). As we identify invariant graphs which belong to the same equivalence class, we have $\varphi_1(\theta) < \varphi_2(\theta) < \varphi_3(\theta)$ m-almost surely. Due to the strict concavity of the fibre maps and the invariance of the three graphs we obtain

$$\log\left(\frac{\varphi_2(\theta+\omega) - \varphi_1(\theta+\omega)}{\varphi_2(\theta) - \varphi_1(\theta)}\right) > \log\left(\frac{\varphi_3(\theta+\omega) - \varphi_2(\theta+\omega)}{\varphi_3(\theta) - \varphi_2(\theta)}\right) \quad m\text{-a.s.} \; . \tag{3.4}$$

However, the following Lemma 3.4 applied to $Y = \mathbb{T}^1$, $S(\theta) = \theta + \omega$, $\nu = m$ and $f = \log(\varphi_{i+1} - \varphi_i)$ $(i = 1, 2)$ yields that the integral with respect to m on both sides equals zero, thus leading to a contradiction. Note that $f \circ S - f$ has the constant majorant $\log(\max_{(\theta,x)\in\mathbb{T}^1\times X} DT_\theta(x))$.

LEMMA 3.4 (Lemma 2 in [**4**]). *Suppose* (Y, S, ν) *is a measure-preserving dynamical system,* $f : Y \to \mathbb{R}$ *is measurable and* $f \circ S - f$ *has an integrable majorant or minorant. Then* $\int_Y f \circ S - f \; d\nu = 0$*.*

For the estimates on the Lyapunov exponents, note that due to the strict concavity there holds

$$\begin{aligned} \lambda(\varphi_1) &= \int_{\mathbb{T}^1} \log\left(\lim_{t\to 0} \frac{T_\theta(\varphi_1(\theta) + t) - \varphi_1(\theta+\omega)}{t}\right) d\theta \\ &> \int_{\mathbb{T}^1} \log\left(\frac{\varphi_2(\theta+\omega) - \varphi_1(\theta+\omega)}{\varphi_2(\theta) - \varphi_1(\theta)}\right) d\theta \; = \; 0 \; . \end{aligned}$$

(The last equality follows again from Lemma 3.4 .) Similarly, we obtain $\lambda(\varphi_2) < 0$.

Now suppose γ^+ is continuous and $T_\theta(\gamma^+(\theta)) \leq \gamma^+(\theta+\omega)\ \forall \theta \in \mathbb{T}^1$. Then we can define a sequence of monotonically decreasing continuous curves by

$$\gamma_n^+(\theta) \ := \ T^n_{\theta-n\omega}(\gamma^+(\theta - n\omega)) \ .$$

As this sequence is bounded below by the invariant graph φ_2 it converges pointwise, and the limit has to be an invariant graph. Since there are no other invariant graphs between φ_2 and γ^+, we must have $\varphi_2 = \lim_{n\to\infty} \gamma_n^+$. Consequently φ_2 is upper semi-continuous as the monotone limit of a sequence of continuous curves. In the same way one can see that φ_1 must be lower semi-continuous.

If φ_1 is not continuous, then the upper bounding graph of the compact invariant set $\overline{\Phi_1}^{ess}$ must be an upper semi-continuous invariant graph which lies between γ^- and γ^+. The only candidate for this is φ_2, such that $\Phi_2 \subseteq \overline{\Phi_1}^{ess}$. However, this is only possible if φ_2 is not continuous. Otherwise, as $\lambda(\varphi_2) < 0$ and due to the Uniform Ergodic Theorem, some iterate of T would act uniformly contracting in the fibres on some neighbourhood U of φ_2. In this case no other invariant graph could intersect U on a set of positive measure, contradicting $\Phi_2 \subseteq \overline{\Phi_1}^{ess}$. Replacing T by T^{-1} we can repeat the same argument for φ_2, such that either both graphs are continuous or both are only semi-continuous and have the same essential closure. This completes the proof.

□

REMARK 3.5. (a) The proof of Theorem 3.2 in [**39**] basically relies on the same idea as the above proof of Theorem 3.3 . It depends on the fact that negative Schwarzian derivative of a $\mathcal{C}^3$-map $F : X \to X$ is equivalent to strictly negative cross ratio distortion. The latter is defined as

$$\mathcal{D}_F(w,x,y,z) \ = \ \frac{\frac{F(y)-F(y)}{y-x} \cdot \frac{F(z)-F(w)}{z-w}}{\frac{F(x)-F(w)}{x-w} \cdot \frac{F(z)-F(y)}{z-y}} \ ,$$

where $w < x < y < z \in X$. Applying the resulting inequality to four invariant graphs and integrating over the circle leads to a contradiction, similar to the argument after (3.4). This excludes the existence of more than three invariant graphs in the situation of Theorem 3.2, and in more or less the same way one obtains the inequalities for the Lyapunov exponents.

(b) We remark that the first part of Theorem 3.2 (meaning statements (i)–(iii)) still holds if the dependence of T_θ on θ is only measurable, provided all other assumptions of the theorem are met and $\theta \mapsto \log(\max_{x\in X} DT_\theta(x))$ has an integrable majorant or minorant. Similarly, in Theorem 3.3 the statement about the number of the invariant graphs and the Lyapunov exponents remain true in the analogous case.

The preceding statements now allow to prove Theorem 2.1:

Proof of Theorem 2.1 . We start with the case where all fibre maps have negative Schwarzian derivative (see (2.5)). Then due to Theorem 3.2, the number of graphs which can exist is at most three . In order to show that the lower bounding graph φ^- is always continuous, let us first collect some facts about the map F: As F has three fixed points, there must exist some $c \in [-2C, 2C]$ with $F''(c) = 0$. However, the negative Schwarzian derivative implies that $F'''(x) < 0$ whenever $F''(x) = 0$ for some $x \in [-2C, 2C]$. Thus there can be only one c with $F''(c) = 0$, and in addition $F''(x)$ will be strictly positive for $x < c$ and strictly negative for $x > c$. Therefore $F_{|[-2C,c)}$ will be strictly convex and $F_{|(c,2C]}$ strictly concave, and this in turn implies that 0 is an unstable fixed point whereas x^- and x^+ are stable. Further $F - \mathrm{Id}$ is strictly positive on $(0, x^+)$ and strictly negative on $(x^-, 0)$, and finally F is a uniform contraction on $[-2C, x^-]$.[3]

As we are in the case of one-sided forcing, for any ϵ with $-\epsilon \in (x^-, 0)$ the set $\mathbb{T}^1 \times [-2C, -\epsilon]$ is mapped into itself, independent of β. Further, as g does not vanish identically, there exist $\epsilon > 0$ and $n \in \mathbb{N}$ such that $T^n(M) \subseteq \mathbb{T}^1 \times [-2C, -\epsilon]$, where $M := \mathbb{T}^1 \times [-2C, 0]$. Consequently

$$\begin{aligned} \bigcap_{n\in\mathbb{N}} T^n(M) \quad &\subseteq \quad \bigcap_{n\in\mathbb{N}} T^n(\mathbb{T}^1 \times [-2C, -\epsilon]) \\ &\subseteq \quad \bigcap_{n\in\mathbb{N}} \mathbb{T}^1 \times [-2C, F^n(-\epsilon)] \;=\; \mathbb{T}^1 \times [-2C, x^-] \;=:\; N \, . \end{aligned}$$

Now T acts uniformly contracting on N in the vertical direction. This means that there will be exactly one invariant graph contained in $N \subseteq M$, which is stable and continuous, and this is of course the lower bounding graph φ^-. In particular $\varphi^- < 0$ independent of β. Furthermore, no other invariant graph can intersect N.

(i) On the one hand, there obviously exist three invariant graphs at $\beta = 0$, namely the constant lines corresponding to the three fixed points. As these are not neutral, they will also persist for small values of β. On the other hand consider $\beta = C$. As we assumed that g takes the maximum value of 1 at least for one $\theta_0 \in \mathbb{T}^1$, the point (θ_0, C) is mapped into M. (Recall that $F : [-2C, 2C] \to [-C, C]$.) But as we have seen, any point in M is attracted to φ^- independent of β. Thus there exists an orbit which starts above the upper bounding graph and ends up converging to φ^-. This means that there can be no other invariant graph apart from φ^-, and as this situation is stable the same will also be true for all β sufficiently close to C.

Consequently, if we define β_0 as the infimum of all $\beta \in (0, C)$ for which there do not exist three continuous invariant graphs, then $\beta_0 \in (0, C)$ and statement (i) holds by definition.

It remains to show that the graphs $\varphi^\pm$ and ψ depend continuously and monotonically on β. Continuity simply follows from the fact that invariant curves with non-zero Lyapunov exponents depend continuously on $\mathcal{C}^1$-distortions of the system. For the monotonicity of φ^+, note that since there is no other invariant graph above, φ^+ is the limit of the iterated upper boundary lines φ_n, which are defined by $\varphi_n(\theta) := T^n_{\theta - n\omega}(2C)$. Due to the one-sided forcing, each of these curves will decrease monotonically as β is increased, and this carries over to φ^+ in the limit. The same argument applies to φ^-, as this is the pointwise limit of the iterated *lower* boundary lines. Finally, note that ψ can be defined as the upper boundary of the set

$$\begin{aligned} &\{(\theta, x) \mid \lim_{n\to\infty} |T^n_\theta(x) - \varphi^-(\theta + n\omega)| = 0\} \\ = \quad &\{(\theta, x) \mid \exists n \in \mathbb{N} : T^n(\theta, x) \in M\} \, . \end{aligned}$$

[3]Note that we do not know whether $c \in (x^-, 0]$, such that this does not imply the second alternative in (2.5).

This set increases with β, and thus the graph ψ will move upwards.

(ii) As all points in M are attracted to φ^-, the two upper invariant graphs for $\beta < \beta_0$ must be contained in M^c. Simply due to continuity, for $\beta \to \beta_0$ the pointwise limits of these curves will be invariant graphs for T_{β_0}, although not necessarily continuous. By compactness, they will be contained in $\overline{M^c}$ and can therefore not coincide with φ^-. Further, they cannot be both distinct and continuous: Due to the non-zero Lyapunov exponents given by Theorem 3.2(iii), this is a stable situation, contradicting the definition of β_0. Thus there only remain the two stated possibilities: Either the two graphs are distinct and not continuous, or they coincide m-a.s. and are neutral (see Theorem 3.2). The compactness of B simply follows from the semi-continuity of the graphs ψ and φ^+.

In the case where ψ equals φ^+ m-a.s., the fact that B is pinched is obvious. Otherwise, it follows from Theorem 3.2 that the two graphs have the same essential closure, which we denote by A. Now all invariant ergodic measures supported on B (namely μ_ψ and μ_{φ^+}) have the same topological closure A, which means that A is minimal and there is no other minimal subset of B. Therefore Theorem 4.6 in [**37**] implies that B is pinched.

(iii) Suppose $\tilde{\beta} = \beta_0 + 2\epsilon$ for any $\epsilon > 0$. We have to show that there is no other invariant graph apart from the lower bounding graph φ^-. For this, it suffices to find an orbit which starts on the upper boundary line and ends up in M: This means that it finally converges to φ^-, which is impossible if there exists another invariant graph above.

First, consider $\beta = \beta_0$ and let θ_1 be chosen such that $\psi(\theta_1) = \varphi^+(\theta_1)$. As the pinched fibres are dense in $\mathbb{T}^1$ and $g(\theta_0) = 1$, we can assume w.l.o.g. that $g(\theta_1 - \omega) \geq \frac{1}{2}$. Further, as the upper boundary lines converge pointwise to φ^+, there exists some $n \in \mathbb{N}$ such that

$$\varphi_n(\theta_1) \ = \ T^n_{\beta_0,\theta_1 - n\omega}(2C) \ \leq \ \varphi^+(\theta_1) + \frac{\epsilon}{2} \ .$$

Now, as the forcing is one-sided (i.e. $g \geq 0$) we have $T^{n-1}_{\tilde{\beta},\theta_1 - n\omega}(2C) \leq T^{n-1}_{\beta_0,\theta_1 - n\omega}(2C)$ and consequently

$$\begin{aligned}
T^n_{\tilde{\beta},\theta_1 - n\omega}(2C) \ &= \ T_{\tilde{\beta},\theta_1 - \omega}(T^{n-1}_{\tilde{\beta},\theta_1 - n\omega}(2C)) \\
&\leq \ T_{\tilde{\beta},\theta_1 - \omega}(T^{n-1}_{\beta_0,\theta_1 - n\omega}(2C)) \\
&= \ F(T^{n-1}_{\beta_0,\theta_1 - n\omega}(2C)) - \tilde{\beta} \cdot g(\theta_1 - \omega) \\
&= \ T^n_{\beta_0,\theta_1 - n\omega}(2C) - (\tilde{\beta} - \beta_0) \cdot g(\theta_1 - \omega) \\
&\leq \ \varphi^+(\theta_1) + \frac{\epsilon}{2} - \epsilon \ < \ \varphi^+(\theta_1) \ = \ \psi(\theta_1) \ .
\end{aligned}$$

However, already for T_{β_0} the orbits of all points below ψ eventually enter M, and again due to the one-sided nature of the forcing this will surely stay true for the respective orbits generated with $T_{\tilde{\beta}}$. Thus, for $\beta = \tilde{\beta}$ the orbit starting at $(\theta_1, 2C)$ ends up in M and therefore converges to the lower bounding graph. As $\epsilon > 0$ was arbitrary, this proves statement (iii).

Now assume the second alternative in (2.5) holds, i.e. for some $c \in (x^-, 0]$ the map $F_{|[c,2C]}$ is strictly concave and $F_{[-2C,x_-]}$ is uniformly contracting. Then the above proof basically remains the same, the only difficulty is to see that for any $\beta \in [0, C]$ there cannot be more than three invariant graphs. However, on the one hand it can be seen as above that the lower bounding graph φ^- is the only invariant graph in M and no other invariant graph intersects M, since all orbits in this set converge to φ^-. On the other hand we can apply Theorem 3.3 with $I(\theta) = [0, 2C]$ to see that there can be at most two invariant graphs in M^c.

Apart from this, the above arguments work in exactly the same way, replacing Theorem 3.2 by Theorem 3.3 where necessary.

□

Proof of Lemma 2.2 . The continuity simply follows from the fact that both the situations above and below the bifurcation are stable, due to the non-zero Lyapunov exponents. Consequently, the sets $\{(\alpha, \beta) \mid \beta < \beta_0(\alpha)\}$ and $\{(\alpha, \beta) \mid \beta > \beta_0(\alpha)\}$ are open, which means that $\alpha \mapsto \beta_0(\alpha)$ must be continuous.

In order to see the monotonicity, let $T_{\alpha,\beta}$ be the system given by (2.1) with $F = F_\alpha$. Suppose that $\tilde{\alpha} > \alpha$. Denote the upper bounding graph of the system $T_{\alpha,\beta_0(\alpha)}$ by φ^+, the invariant graph in the middle by ψ. As all points on or below the 0-line eventually converge to the lower bounding graph (see the proof of Theorem 2.1), the invariant graphs ψ and φ^+ must be strictly positive. As ψ is lower semi-continuous and $\varphi^+ \geq \psi$, both graphs are uniformly bounded away from 0. Thus, there exists some $\delta > 0$ such that $\delta \leq \varphi^+ \leq 1 - \delta$.

For any $x \in [\delta, 1 - \delta]$ the map $F_\alpha(x)$ is strictly increasing in α.[4] Due to compactness this means that there exists $\epsilon > 0$, such that $F_{\tilde{\alpha}} > F_\alpha + \epsilon$ on $[\delta, 1 - \delta]$. Let $\tilde{\beta} := \beta_0(\alpha) + \epsilon$. Then

$$T_{\tilde{\alpha},\tilde{\beta},\theta}(x) > T_{\alpha,\beta_0(\alpha),\theta}(x) \quad \forall (\theta, x) \in \mathbb{T}^1 \times [\delta, 1 - \delta] .$$

Consequently $T_{\tilde{\alpha},\tilde{\beta}}$ maps the graph φ^+ strictly above itself, which means that the upper bounding graph $\tilde{\varphi}^+$ of this system must be above φ^+. It can therefore not coincide with the lower bounding graph, which lies below the 0-line. Hence $\beta_0(\tilde{\alpha}) \geq \tilde{\beta} > \beta_0(\alpha)$.

□

3.3. Sink-source-orbits and SNA: Proof of Theorem 2.4

Suppose that T satisfies the assumptions of Theorem 2.4 and denote the upper and lower bounding graph by φ^+ and φ^-, respectively. Suppose there exists no non-continuous invariant graph with negative Lyapunov exponent, but a point $(\theta_0, x_0) \in \mathbb{T}^1 \times X$ with $\lambda^+(\theta_0, x_0) > 0$ and $\lambda^-(\theta_0, x_0) > 0$ (i.e. a sink-source-orbit). Let

$$\psi^+(\theta) := \inf\{\varphi(\theta) \mid \varphi \text{ is a continuous } T\textit{-invariant graph with } \varphi(\theta_0) \geq x_0\} ,$$

with $\psi^+ :\equiv \varphi^+$ if no such graph φ exists. Similarly, define

$$\psi^-(\theta) := \sup\{\varphi(\theta) \mid \varphi \text{ is a continuous } T\textit{-invariant graph with } \varphi(\theta_0) \leq x_0\} ,$$

[4] We have

$$\frac{\partial}{\partial\alpha} F_\alpha(x) = \frac{\partial}{\partial\alpha}\left(\frac{\arctan(\alpha x)}{\arctan(\alpha)}\right) = \left(\frac{x \cdot \arctan(\alpha)}{1 + \alpha^2 x^2} - \frac{\arctan(\alpha x)}{1 + \alpha^2}\right) \cdot \arctan(\alpha)^{-2} .$$

This is positive if and only if

$$G_\alpha(x) := x \cdot \arctan(\alpha) \cdot (1 + \alpha^2) - \arctan(\alpha x) \cdot (1 + \alpha^2 x^2)$$

is positive. However, it is easy to verify that $G_\alpha(0) = G_\alpha(1) = 0$ and G_α is strictly concave on $[0, 1]$, i.e. $\frac{\partial^2}{\partial^2 x} G_\alpha(x) < 0 \ \forall x \in [0, 1]$, such that $G_\alpha(x) > 0 \ \forall x \in (0, 1)$.

with $\psi^- :\equiv \varphi^-$ if there is no such graph φ. By the continuity and monotonicity of the fibre maps, ψ^+ and ψ^- will be invariant graphs again. In addition, ψ^+ will be upper and ψ^- lower semi-continuous and $\psi^- \leq \psi^+$. Thus, the set $A := [\psi^-, \psi^+]$ is compact. By a semi-uniform ergodic theorem contained in [**25**] (Theorem 1.9), both $\lambda^+(\theta_0, x_0)$ and $-\lambda^-(\theta_0, x_0)$ must be contained in the convex hull of the set

$$\left\{ \int_A \log DT_\theta(x)\, d\mu(\theta, x) \mid \mu \textit{ is a } T_{|A}\textit{-invariant and ergodic probability measure} \right\} .$$

As all ergodic measures are associated to invariant graphs (see (1.5)), this means that there must exist invariant graphs with positive and negative Lyapunov exponents in A. However, as we assumed that all stable invariant graphs are continuous and there are no continuous invariant graphs contained in the interior of A by the definition of $\psi^\pm$, the only possible candidates for a negative Lyapunov exponent are ψ^+ and ψ^-. We consider the case where only $\lambda(\psi^-) < 0$, if ψ^+ or both invariant graphs are stable this can be dealt with similarly. Note that by the assumption we made at the beginning, the negative Lyapunov exponent ensures that ψ^- must be continuous.

Consequently, the convergence of the Lyapunov exponents is uniform on ψ^-, such that there there is and open neighbourhood of this curve which is uniformly contracted in the vertical direction by some iterate of T. Therefore, if we define

$$\tilde{\psi}^-(\theta) \;:=\; \inf\{x \geq \psi^-(\theta) \mid \limsup_{n\to\infty} |T^n_\theta(x) - \psi^-(\theta + n\omega)| > 0\} .$$

then $\tilde{\psi}^- > \psi^-$, and in addition $\tilde{\psi}^-$ is lower semi-continuous. Note that

$$\lim_{n\to\infty} |T^n_\theta(x) - \psi^-(\theta + n\omega)| \;=\; 0 \qquad \forall(\theta, x) \in [\psi^-, \tilde{\psi}^-)$$

by definition. The forward orbit of (θ_0, x_0) cannot converge to ψ^- as this contradicts $\lambda^+(\theta_0, x_0) > 0$. Therefore $x_0 \geq \tilde{\psi}^-(\theta_0)$. Further, there holds $\tilde{\psi}^- \leq \psi^+$. This means that (θ_0, x_0) is contained in the compact set $\tilde{A} := [\tilde{\psi}^-, \psi^+]$. But as $\tilde{A}$ does not contain an invariant graph with negative Lyapunov exponent anymore, this contradicts $\lambda^-(\theta_0, x_0) > 0$, again by Theorem 1.9 in [**25**].

The existence of a strange non-chaotic repeller follows in the same way by regarding the inverse of T restricted to the global attractor.

□

CHAPTER 4

The strategy for the construction of the sink-source-orbits

The inductive construction of longer and longer trajectories which are expanding in the forwards and contracting in the backwards direction (compare Lemma 2.6) will be a rather complicated inductive procedure. On the one hand, a substantial amount of effort will have to be put into introducing the right objects and providing a number of preliminary estimates and technical statements in Section 5. On the other hand, it will sometimes be quite hard to see the motivation for all this until the actual construction is carried out in Section 6. In order to give some guidance to the reader in the meanwhile, we will try to sketch a rough outline of the overall strategy in this section, and discuss at least some of the main problems we will encounter. In particular, we will try to indicate how a recursive structure appears in the construction, induced by the recurrence behaviour of the underlying irrational rotation.

To this end, we will start by deriving some first (easy) estimates, which will make it much easier to talk about what happens further. This will show that up to a certain point the construction is absolutely straightforward. The further strategy will then only be outlined, as the tools developed in Section 5 are needed before it can finally be converted into a rigorous proof in Section 6.

4.1. The first stage of the construction

As mentioned in Section 1.4, for a suitable choice of the functions F and g in (2.1) we can expect that the tips of the peaks correspond to a sink-source-orbit. However, as we do not know the bifurcation parameter exactly, we can only approximate it and show that in each step of the approximation there is a longer finite trajectory with the required behaviour. The existence of the sink-source-orbit at the bifurcation point will then follow from Lemma 2.6 .

As we will concentrate only on trajectories in the orbit of the 0-fibre, the following notation will be very convenient:

DEFINITION 4.1. *For the map T_β defined in Theorem 2.7 with fibre maps $T_{\beta,\theta}$, let*

$$T_{\beta,\theta,n} := T_{\beta,\theta+\omega_{n-1}} \circ \ldots \circ T_{\beta,\theta}$$

if $n > 0$ and $T_{\beta,\theta,0} := Id$. Further, for any pair $l \leq n$ of integers let

$$\xi_n(\beta, l) := T_{\beta,\omega_{-l},n+l}(3) \ .$$

In other words, $\xi_n(\beta, l)$ is the x-value of that point from the T_β-forward orbit of $(\omega_{-l}, 3)$, which lies on the ω_n-fibre. Thus, the lower index always indicates the fibre on which the respective point is located.

Slightly abusing language, we will refer to $(\xi_j(\beta,l))_{n\geq -l}$ as the forward orbit of the point $(\omega_{-l},3)$, suppressing the θ-coordinates.

Note that under the assumptions of Theorem 2.7 (which imply in particular that we are in the case of one-sided forcing, i.e. $g \geq 0$) the mapping $\beta \mapsto \xi_n(\beta,l)$ is monotonically decreasing for any fixed numbers l and n, with strict monotonicity if $l \geq 0$ and $n \geq 1$ since $g(0) = 1$. In addition, we claim that when $n \geq 1$ and $l \geq 0$, the interval $\overline{B_{\frac{1}{\alpha}}(0)}$ is covered as β increases from 0 to $\frac{3}{2}$, i.e.

$$\xi_n(\tfrac{3}{2},l) \ < \ -\tfrac{1}{\alpha} \ . \tag{4.1}$$

In order to see this, note that $\xi_0(\beta,l)$ is always smaller than 3, such that $\xi_0(\beta,l) - x_\alpha \leq 2 - \frac{2}{\sqrt{\alpha}}$. Therefore, using $F(x_\alpha) = x_\alpha$, (2.7) and $g(0) = 1$ we obtain

$$\xi_1(\tfrac{3}{2},l) \ = \ F(\xi_0(\beta,l)) - \tfrac{3}{2}\cdot g(0) \ \leq \ x_\alpha + \frac{2-\frac{2}{\sqrt{\alpha}}}{2\sqrt{\alpha}} - \frac{3}{2} \ = \ \frac{3}{\sqrt{\alpha}} - \frac{1}{\alpha} - \frac{1}{2} \ .$$

By (2.2) the right side is smaller than $-\frac{1}{\alpha}$, and as $\mathbb{T}^1 \times [-3, -\frac{1}{\alpha})$ is always mapped into itself this proves our claim.

From now on, we use the following notation: For any pair k,n of integers with $k \leq n$ let

$$[k,n] \ := \ \{k,\ldots,n\} \ . \tag{4.2}$$

What we want to derive is a statement of the following kind

> *If $\xi_N(\beta,l) \in \overline{B_{\frac{1}{\alpha}}(0)}$ for 'suitable' integers $l \leq 0$ and $N \geq 1$, then $\xi_j(\beta,l) \in \overline{B_{\frac{1}{\alpha}}(0)}$ for 'most' $j \in [1,N]$ and $\xi_j(\beta,l) \geq \gamma$ for 'most' $j \in [-l,0]$.*

Of course, we have to specify what 'suitable' and 'most' mean, but as this will be rather complicated we postpone it for a while. As (4.1) implies that there always exist values of $\beta \in [0,\frac{3}{2}]$ with $\xi_n(\beta,l) \in \overline{B_{\frac{1}{\alpha}}(0)}$, such a statement would ensure the existence of trajectories which spend most of the backward time in the contracting region and most of the forward time in the expanding region. This is exactly what is needed for the application of Lemma 2.6 . As mentioned, up to a certain point things are quite straightforward:

LEMMA 4.2. *Suppose that the assumptions of Theorem 2.7 hold. Further, let $n \geq 1$, $l \geq 0$ and assume that*

$$d(\omega_j,0) \ \geq \ \frac{3\gamma}{L_2} \quad \forall j \in [-l,-1] \cup [1,n-1] \ .$$

Then $\xi_n(\beta,l) \in \overline{B_{\frac{1}{\alpha}}(0)}$ implies $\beta \in [1+\frac{1}{\sqrt{\alpha}}, 1+\frac{3}{\sqrt{\alpha}}]$,

$$\xi_j(\beta,l) \ \in \ \overline{B_{\frac{1}{\alpha}}(0)} \qquad \forall j \in [1,n] \tag{4.3}$$

and

$$\xi_j(\beta,l) \ \geq \ \gamma \qquad \forall j \in [-l,0] \ . \tag{4.4}$$

The proof relies on the following basic estimate:

LEMMA 4.3. *Suppose that the assumptions of Theorem 2.7 hold. Further, assume that* $\beta \le 1+\frac{4}{\sqrt{\alpha}}$, $j \ge -l$ *and* $d(\omega_j,0) \ge \frac{3\gamma}{L_2}$. *Then* $\xi_j(\beta,l) \ge \frac{1}{\alpha}$ *implies* $\xi_{j+1}(\beta,l) \ge \gamma$ *and* $\xi_j(\beta,l) \le -\frac{1}{\alpha}$ *implies* $\xi_{j+1}(\beta,l) \le -\gamma$. *Consequently,* $\xi_{j+1}(\beta,l) \in \overline{B_{\frac{1}{\alpha}}(0)}$ *implies* $\xi_j(\beta,l) \in \overline{B_{\frac{1}{\alpha}}(0)}$.

PROOF. Suppose that $\xi_j(\beta,l) \ge \frac{1}{\alpha}$. Using $d(\omega_j,0) \ge \frac{3\gamma}{L_2}$ and (2.11) we obtain that $g(\omega_j) \le 1-3\gamma$. Therefore

$$\begin{aligned}\xi_{j+1}(\beta,l) &= F(\xi_j(\beta,l)) - \beta\cdot g(\omega_j)\\ &\overset{(2.8)}{\ge} 1-\gamma-(1+\tfrac{4}{\sqrt{\alpha}})(1-3\gamma) \;\ge\; 2\gamma-\tfrac{4}{\sqrt{\alpha}} \;\overset{(2.2)}{\ge}\; \gamma\,.\end{aligned}$$

As $g \ge 0$, we also see that $\xi_j(\beta,l) \le -\frac{1}{\alpha}$ implies

$$\xi_{j+1}(\beta,l) \;\le\; F(\xi_j(\beta,l)) \;\overset{(2.8)}{\le}\; -(1-\gamma) \;\overset{(2.1)}{\le}\; -\gamma\,.$$

□

Proof of Lemma 4.2:
Suppose that $\xi_n(\beta,l) \in \overline{B_{\frac{1}{\alpha}}(0)}$. We first show that $\beta \le 1+\frac{3}{\sqrt{\alpha}}$: As $\xi_0(\beta,l) \le 3$ we can use $F(x_\alpha)=x_\alpha$ and (2.7) to see that $F(\xi_0(\beta,l)) \le 1+\frac{3}{\sqrt{\alpha}}-\frac{1}{\alpha}$. As $g(0)=1$ this gives

$$\xi_1(\beta,l) \;=\; F(\xi_0(\beta,l))-\beta \;\le\; \left(1+\frac{3}{\sqrt{\alpha}}-\beta\right)-\frac{1}{\alpha}\,.$$

Thus, for $\beta > 1+\frac{3}{\sqrt{\alpha}}$ we have $\xi_1(\beta,l) < -\frac{1}{\alpha}$, and as $\mathbb{T}^1 \times [-3,-\frac{1}{\alpha})$ is mapped into itself this would yield $\xi_n(\beta,l) < -\frac{1}{\alpha}$, contradicting our assumption. Therefore $\xi_n(\beta,l) \in \overline{B_{\frac{1}{\alpha}}(0)}$ implies $\beta \le 1+\frac{3}{\sqrt{\alpha}}$.

Now we can apply Lemma 4.3 to all $j \in [1,n-1]$ and obtain $\xi_j(\beta,l) \in \overline{B_{\frac{1}{\alpha}}(0)}$ $\forall j \in [1,n]$ by backwards induction on j, starting at $j=n$. Similarly, $\xi_j(\beta,l) \ge \gamma$ $\forall j=-l,\dots,0$ follows from $\xi_{-l}(\beta,l)=3 \ge \gamma$ by forwards induction, as we can again apply Lemma 4.3 to all $j \in [-l,-1]$.

It remains to prove that $\beta \ge 1+\frac{1}{\sqrt{\alpha}}$. We already showed that $\xi_0(\beta,l) \ge \gamma \ge x_\alpha - 1$, such that we can use $F(x_\alpha)=x_\alpha$ and (2.7) again to see that

$$\xi_1(\beta,l) \;\ge\; x_\alpha - \frac{1}{2\sqrt{\alpha}} - \beta \;=\; 1+\frac{3}{2\sqrt{\alpha}}-\beta \;\overset{(2.2)}{\ge}\; \left(1+\frac{1}{\sqrt{\alpha}}-\beta\right)+\frac{1}{\alpha}\,.$$

As we also showed above that $\xi_1(\beta,l) \le \frac{1}{\alpha}$, the required estimate follows.

□

4.2. Dealing with the first close return

As we have seen above, everything works fine as long as the ω_j do not enter the interval $B_{\frac{3\gamma}{L_2}}(0)$ again. Thus, in the context of Section 1.4 the critical region $\mathcal{C}$ corresponds to the vertical strip $B_{\frac{3\gamma}{L_2}}(0) \times [-3,3]$. We will now sketch the argument by which the construction can be continued even beyond the first return to this critical region:

Suppose $m \in \mathbb{N}$ is the first time such that $d(\omega_m,0) < \frac{3\gamma}{L_2}$ and fix some $l \le m-1$. Then Lemma 4.2 yields information up to time m, meaning that we can apply it whenever $n \le m$. But we cannot ensure that $\xi_{m+1}(\beta,l) \in \overline{B_{\frac{1}{\alpha}}(0)}$ implies

$\xi_m(\beta,l) \in \overline{B_{\frac{1}{\alpha}}(0)}$ as before. In fact, this will surely be wrong when ω_m is too close to 0, such that $g(\omega_m) \approx 1$. In order to deal with this, we will define a certain 'exceptional' interval $J(m) = [m-l^-,\ldots,m+l^+]$. The integers l^- and l^+ will have to be chosen very carefully later on, but for now the reader should just assume that they are quite small in comparison to both m and l. Then, instead of showing that $\xi_{m+1}(\beta,l) \in \overline{B_{\frac{1}{\alpha}}(0)}$ implies $\xi_m(\beta,l) \in \overline{B_{\frac{1}{\alpha}}(0)}$ as before, we will prove that

$$\xi_{m+l^++1}(\beta,l) \in \overline{B_{\frac{1}{\alpha}}(0)} \quad \textit{implies} \quad \xi_{m-l^--1}(\beta,l) \in \overline{B_{\frac{1}{\alpha}}(0)} \;. \tag{4.1}$$

Using Lemma 4.2, the latter then ensures that $\xi_j(\beta,l) \in \overline{B_{\frac{1}{\alpha}}(0)}\ \forall j \in [1, m-l^--1]$.

Recall that as we are in the case of one-sided forcing, the dependence of $\xi_n(\beta,l)$ on β is strictly monotone. Thus, in order to prove (4.1), it will suffice to consider the two unique parameters β^+ and β^- which satisfy

$$\xi_{m-l^--1}(\beta^+,l) \;=\; \frac{1}{\alpha} \tag{4.2}$$

and

$$\xi_{m-l^--1}(\beta^-,l) \;=\; -\frac{1}{\alpha} \;. \tag{4.3}$$

If we can then show the two inequalities

$$\xi_{m+l^++1}(\beta^+,l) \;>\; \frac{1}{\alpha} \tag{4.4}$$

and

$$\xi_{m+l^++1}(\beta^-,l) \;<\; -\frac{1}{\alpha} \;, \tag{4.5}$$

this immediately implies (4.1).

Now, first of all the fact that (4.5) follows from (4.3) is obvious, as $\mathbb{T}^1 \times [-3, -\frac{1}{\alpha}]$ is mapped into $\mathbb{T}^1 \times [-3, -(1-\gamma)]$ by (2.8), independent of the parameter β. Thus, it remains to show (4.4). This will be done by comparing the orbit[1]

$$\xi_{m-l^--1}(\beta^+,l),\ldots,\xi_{m+l^++1}(\beta^+,l) \tag{4.6}$$

with suitable *'reference orbits'*, on which information is already available by Lemma 4.2 . In order to make such comparison arguments precise (as sketched in Figure 4.1 below), we will need the following concept:

DEFINITION 4.4. *For any $\beta_1,\beta_2 \in [0,\frac{3}{2}]$ and $\theta_1,\theta_2 \in \mathbb{T}^1$, the* ***error term*** *is defined as*

$$\mathrm{err}(\beta_1,\beta_2,\theta_1,\theta_2) \;:=\; \sup_{n\in\mathbb{Z}} |\beta_1 \cdot g(\theta_1+\omega_n) - \beta_2 \cdot g(\theta_2+\omega_n)| \;.$$

Note that $\mathrm{err}(\beta_1,\beta_2,\theta_1,\theta_2) = \sup_{n\in\mathbb{Z}} \parallel T_{\beta_1,\theta_1+\omega_n} - T_{\beta_2,\theta_2+\omega_n} \parallel_\infty$.

The next remark gives a basic estimate:

REMARK 4.5. Suppose that g has Lipschitz-constant L_1 (as in (2.10)). Further, assume that $\theta_1 = \omega_k$, $\theta_2 = \omega_{k+m}$ for some $k,m \in \mathbb{Z}$, $d(\omega_m,0) \leq \frac{2\epsilon}{L_2}$, and $\beta_1,\beta_2 \in [1,\frac{3}{2}]$ satisfy $|\beta_1-\beta_2| < 2\epsilon$. Then

$$\mathrm{err}(\beta_1,\beta_2,\theta_1,\theta_2) \leq K \cdot \epsilon$$

where $K := 3 \cdot \frac{L_1}{L_2} + 2$.

[1] Recall that we suppress the θ-coordinate ω_j of points $(\omega_j, \xi_j(\beta,l))$ from the forward orbit of $(\omega_{-l},3)$.

PROOF. For any $n \in \mathbb{N}$, let $j := k + n$. Then $\omega_k + \omega_n = \omega_j$ and $\omega_{k+m} + \omega_n = \omega_{j+m}$. Thus, the above estimate follows from

$$|\beta_1 \cdot g(\omega_j) - \beta_2 \cdot g(\omega_{j+m})| \leq$$

$$\beta_1 \cdot |g(\omega_j) - g(\omega_{j+m})| + g(\omega_{j+m}) \cdot |\beta_1 - \beta_2| \;\leq\; \beta_1 \cdot \frac{2\epsilon}{L_2} \cdot L_1 + 2\epsilon \;\leq K \cdot \epsilon$$

□

Thus, even if two finite trajectories are generated with slightly different parameters and are not located on the same but only on nearby fibres, the fibre maps which produce them will still be almost the same. This makes it possible to compare two such orbits, at least up to a certain extent. For the remainder of this section, the reader should just assume that the remaining differences between the fibre maps can always be neglected. Of course, when the construction is made rigorous later on it will be a main issue to show that this is indeed the case.

Let us now turn to Figure 4.1, which illustrates the argument used to derive (4.4). The first reference orbit, shown as crosses, is generated with the unique parameter β^* that satisfies $\xi_m(\beta^*, l) = 0$. Due to Lemma 4.2 (with $n = m$), we know that this orbit always stays in the expanding region before, i.e.

$$\xi_j(\beta^*, l) \;\in\; \overline{B_{\frac{1}{\alpha}}(0)} \quad \forall j = 1, \ldots, m-1\;. \tag{4.7}$$

Recall that β^+ was defined by $\xi_{m-l^--1}(\beta^+, l) = \frac{1}{\alpha}$. This implies $\xi_{m-l^-}(\beta^+, l) \geq \gamma$ by Lemma 4.3 . Thus, the 'new' orbit $\xi_{m-l^--1}(\beta^+, l), \ldots, \xi_{m+l^++1}(\beta^+, l)$ (corresponding to the black squares in Figure 4.1) leaves the expanding region and enters the contracting region (A), whereas the reference orbit (crosses) stays in the expanding region at the same time, i.e. $\xi_{m-l^-}(\beta^*, l) \in \overline{B_{\frac{1}{\alpha}}(0)}$, by (4.7). Afterwards, due to the strong expansion on $\mathbb{T}^1 \times \overline{B_{\frac{2}{\alpha}}(0)}$ it is not possible for the new orbit to approach the reference orbit anymore, such that it will stay 'trapped' in the contracting region (B). In this way, we will obtain[2]

$$\xi_j(\beta^+, l) \;\geq\; \gamma \quad \forall j = m - l^-, \ldots, m\;. \tag{4.8}$$

Now we start to use a second reference orbit, namely $\xi_{-l^-}(\beta^+, l), \ldots, \xi_{l^++1}(\beta^+, l)$, shown by the circles in Figure 1.8. Note that this time it will be generated with exactly the same parameter β^+ as the new orbit, but located on slightly different fibres. By Lemma 4.2 (with $n = m - l^- - 1$, note that $\xi_{m-l^--1}(\beta^+, l) \in \overline{B_{\frac{1}{\alpha}}(0)}$ by definition), we know that

$$\xi_j(\beta^+, l) \;\geq\; \gamma \quad \forall j = -l^-, \ldots, 0 \tag{4.9}$$

and

$$\xi_j(\beta^+, l) \;\in\; \overline{B_{\frac{1}{\alpha}}(0)} \quad \forall j = 1, \ldots, l^+ + 1\;. \tag{4.10}$$

Combining (4.8) and (4.9), we see that the two orbits we want to compare both spend the first l^- iterates in the contracting region. Thus they are attracted to each other, and consequently $|\xi_0(\beta^+, l) - \xi_m(\beta^+, l)|$ will be very small (C). In fact, if l^-

[2]We should mention that in this particular situation (4.8) could still be derived directly from Lemma 4.3 . However, the advantage of the described comparison argument is that it is more flexible and will also work for later stages of the construction.

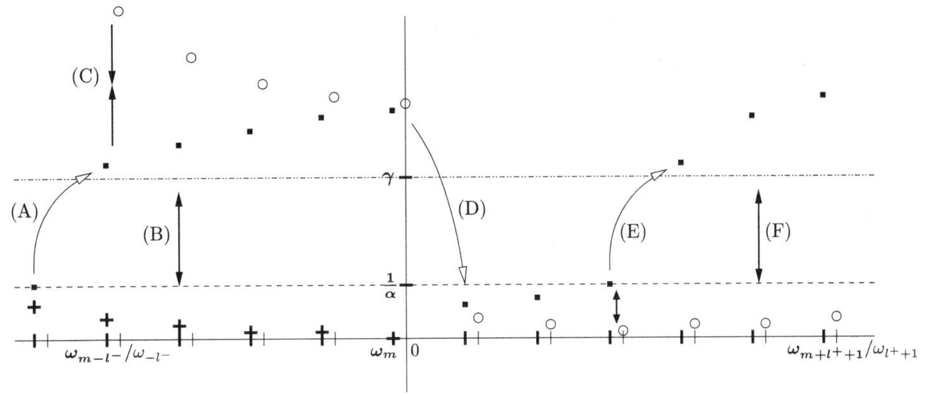

FIGURE 4.1. The above diagram shows three finite trajectories: The 'new' orbit $\xi_{m-l^--1}(\beta^+,l), \ldots, \xi_{m+l^++1}(\beta^+,l)$ (black squares), the first reference orbit $\xi_{m-l^--1}(\beta^*,l), \ldots, \xi_m(\beta^*,l)$ (crosses) and the second reference orbit $\xi_{-l^-}(\beta^+,l), \ldots, \xi_{l^++1}(\beta^+,l)$ (circles). For convenience, successive iterates on the circle are drawn in straight order. (This corresponds to the situation where either the rotation number ω is very small, or where we consider a q-fold cover of the circle $\mathbb{T}^1$.) After the first iterate, the new orbit leaves the expanding and enters the contracting region (A). Afterwards, the first reference orbit together with the strong expansion on $\mathbb{T}^1 \times \overline{B_{\frac{2}{\alpha}}(0)}$ ensure that the new orbit stays in the contracting region as the ω_m-fibre is approached (B). Consequently, it gets attracted to the second reference orbit, which also lies in the contracting region (C). When the 0-fibre is passed, the forcing acts stronger on the second reference orbit (which passes exactly through the 0-fibre) than on the new orbit (which only passes through the ω_m-fibre). Therefore, the new orbit will be slightly above the second reference orbit afterwards (D). From now on, the expansion on $\mathbb{T}^1 \times \overline{B_{\frac{2}{\alpha}}(0)}$ ensures that the new orbit eventually gets pushed out of the expanding region (E), and stays in the contracting region afterwards (F).

has been chosen large enough, then this difference will be of the same magnitude as $\epsilon := L_2 \cdot d(\omega_m, 0)$, i.e.

$$|\xi_0(\beta^+,l) - \xi_m(\beta^+,l)| \leq \kappa \cdot \epsilon \tag{4.11}$$

for a suitable constant $\kappa > 0$.

The next step is crucial: When going from $\xi_0(\beta^+,l)$ to $\xi_1(\beta^+,l)$, the downward forcing takes its maximum (i.e. $g(0) = 1$). In contrast to this, in the transition from $\xi_m(\beta^+,l)$ to $\xi_{m+1}(\beta^+,l)$ the forcing function $g(\omega_m)$ is only close to 1. More precisely, (2.11) yields $g(\omega_m) \leq 1 - \epsilon$. Therefore

$$\xi_{m+1}(\beta^+,l) - \xi_1(\beta^+,l) \geq$$
$$\geq \beta^+ \cdot \epsilon - |F(\xi_m(\beta^+,l)) - F(\xi_0(\beta^+,l))| \overset{(2.7)}{\geq} \epsilon - \frac{\kappa \cdot \epsilon}{\sqrt{\alpha}} \geq \frac{\epsilon}{2},$$

where we have assumed that $\sqrt{\alpha}$ will be larger than 2κ and $\beta^+ \geq 1$. Thus, when the orbits pass the 0- and ω_m-fibre, respectively, a difference is created and the new orbit will be slightly above the reference orbit afterwards (D). But from that point on, the reference orbit stays in the expanding region by (4.10). Therefore, the small difference will be expanded until finally the new orbit is 'thrown out' upwards (E) and gets trapped in the contracting region again (F). This will complete the proof of (4.4).

The crucial point now is the fact that the scheme in Figure 4.1 offers a lot of flexibility. We have described the argument for the particular case of the first close return, but in fact all close returns will be treated in a similar way. The only difference will be the fact that the reference orbits we use in the later stages of the construction may not stay in the expanding (or respectively contracting) region all of the considered times. However, this will still be true for most times, and that is sufficient to ensure that on average the expansion (or contraction) overweights and the new orbit shows the required behaviour.

4.3. Admissible and regular times

The picture we have drawn so far is already sufficient to motivate some further terminology. As we have seen above, not all times $N \in \mathbb{N}$ are suitable for the construction, in the sense of the statement given below (4.2). Thus, we will distinguish between times which are *'admissible'* and others which are not. Only for admissible N we will show that $\xi_N(\beta, l) \in \overline{B_{\frac{1}{\alpha}}(0)}$ allows to draw conclusions about previous times $j < N$. To be more precise, for any given admissible N we will define a set $R_N \subseteq [1, N]$ and show that $\xi_N(\beta, l) \in \overline{B_{\frac{1}{\alpha}}(0)}$ implies $\xi_j(\beta, l) \in \overline{B_{\frac{1}{\alpha}}(0)}$ $\forall j \in R_N$. The integers $j \in R_N$ will then be called *'regular with respect to N'*. The precise definitions of admissible and regular times will be given in Sections 5.3 and 5.4 .

In order to give an example, consider the situation of the previous section: There, all points $N \leq m$ are admissible, and so is $m + l^+ + 1$, but $m+1, \ldots, m+l^+$ are not admissible. Further, for any $N \leq m$ we can choose $R_N = [1, N]$, and the set R_{m+l^++1} contains at least all points from $[1, m + l^+ + 1] \setminus J(m)$. However, it will turn out that we have to define even more times as regular w.r.t. $m + l^+ + 1$, and thus derive information about them, as this will be needed in the later stages of the construction. Namely, the additional points we need to be regular are $m + 1, \ldots, m + l^+$. The reason why this is necessary is explained in Section 4.4 and Figure 4.2. However, in this particular situation it is not difficult to achieve this:

As ω_m is a close return, we can expect (and also ensure by using the diophantine condition and suitable assumptions on γ) that $\omega_{m+1}, \ldots, \omega_{m+l^+}$ are rather far away from 0, in particular not contained in $B_{\frac{3\gamma}{L_2}}(0)$. But this means that we can apply Lemma 4.3 to $m + 1, \ldots, m + l^+$ and obtain that $\xi_{m+l^++1}(\beta, l) \in \overline{B_{\frac{1}{\alpha}}(0)}$ implies $\xi_j(\beta, l) \in \overline{B_{\frac{1}{\alpha}}(0)}$ $\forall j = m + 1, \ldots, m + l^+$ by backwards induction on j. Thus, if we divide the interval $J(m)$ into two parts $J^-(m) := [m - l^-, m]$ and $J^+(m) := [m+1, m+l^+]$, then we can also define all points in the right part $J^+(m)$ as regular, such that $R_{m+l^++1} = [1, m + l^+ + 1] \setminus J^-(m)$.

The reader should keep in mind that although most points will be both regular and admissible, the difference between the two notions is absolutely crucial. For example, for the argument in the previous section it was vitally important that m itself is admissible, as the first reference orbit ended exactly on the ω_m-fibre. But on the other hand, m will not be regular w.r.t. any $N \geq m$, as it is a close return itself and certainly contained in $J^-(m)$.

4.4. Outline of the further strategy

For a certain while the arguments from Section 4.2 will allow to continue the construction as described. When there is another close return at time $m' > m$ and $d(\omega_{m'}, 0)$ is approximately of the same size as $d(\omega_m, 0)$, then the diophantine condition will ensure that m and m' are far apart. Thus, if we define an exceptional interval $J(m')$ again, this will be far away from $J(m)$ and we can proceed more or less as before. However, we have also seen that the minimal lengths of l^- and l^+ depend on how close ω_m is to 0, as there must be enough time for the contraction to work until (4.11) is ensured, and similarly for the expansion until the new orbit is pushed out of the expanding region. To be more precise, let $p \in \mathbb{N}_0$ such that $\epsilon = L_2 \cdot d(\omega_m, 0) \in [\alpha^{-(p+1)}, \alpha^{-p})$. Then the minimal lengths of l^- and l^+ will depend linearly on p, as the expansion and contraction rates are always between $\alpha^{\pm\frac{1}{2}}$ and $\alpha^{\pm 2}$ by (2.6) and (2.7). Thus, at some stage we will encounter a close return at time $\hat{m}$, for which the quantities $\hat{l}^-$ and $\hat{l}^+$ needed to define a suitable interval $J(\hat{m}) = [\hat{m} - \hat{l}^-, \hat{m} + \hat{l}^+]$ are larger than l and m.

At first, assume that only $\hat{l}^+ > m$, whereas $\hat{l}^-$ is still smaller that l. As mentioned, we will be able to show that

$$(4.1) \qquad \xi_{\hat{m}+\hat{l}^++1}(\beta, l) \in \overline{B_{\frac{1}{\alpha}}(0)} \quad \textit{implies} \quad \xi_{\hat{m}-\hat{l}^--1}(\beta, l) \in \overline{B_{\frac{1}{\alpha}}(0)}$$

by a slight modification of the argument sketched in Figure 4.1 . In fact, for the left side there is no difference: If β^+ and β^* are again chosen such that $\xi_{\hat{m}-\hat{l}^--1}(\beta^+, l) = \frac{1}{\alpha}$ and $\xi_{\hat{m}}(\beta^*, l) = 0$, then the first reference orbit $\xi_{\hat{m}-\hat{l}^--1}(\beta^*, l), \ldots, \xi_{\hat{m}}(\beta^*, l)$ will again stay in the expanding region all the time. Therefore we can use it to control the first part $\xi_{\hat{m}-\hat{l}^--1}(\beta^+, l), \ldots, \xi_{\hat{m}}(\beta^+, l)$ of the new orbit as before, and conclude that it always stays in the contraction region. As the same will be true for the first part $\xi_{-\hat{l}^-}(\beta^+, l), \ldots, \xi_0(\beta^+, l)$ of the second reference orbit, the contraction ensures again that $|\xi_{\hat{m}}(\beta^+, l) - \xi_0(\beta^+, l)|$ is small enough (compare (4.11)), and consequently $\xi_{\hat{m}+1}(\beta^+, l)$ will be slightly above $\xi_1(\beta^+, l)$ after the 0-fibre is passed (compare (4.12)).

But afterwards, the second part $\xi_1(\beta^+, l), \ldots, \xi_{\hat{l}^++1}(\beta^+, l)$ of the reference orbit will not stay in the expanding region all the time, as the exceptional interval $J(m)$ is contained in $[1, \hat{l}^+]$ and the points in $J^-(m)$ will not be regular w.r.t. $\hat{m} - \hat{l}^- - 1$. However, as all other points in $[1, \hat{l}^+]$ are regular, it is still possible to show that the new orbit is eventually pushed out of the expanding region again, but this needs a little bit more care than before. Figure 4.2 shows one of the problems we will encounter, and thereby explains why it is so vitally important that we have information about the points in $J^+(m)$ as well, i.e. define them as regular before.

Now, we can begin to see how a recursive structure in the definition of the sets R_N appears: In order to have enough information for even later stages in the construction, we will again have to define at least most points in $J^+(\hat{m}) = [\hat{m} + 1, \hat{m} + \hat{l}^+]$ as regular. As it will turn out, we will be able to show that $\xi_{\hat{m}+\hat{l}^++1} \in \overline{B_{\frac{1}{\alpha}}(0)}$ implies $\xi_{\hat{m}+j}(\beta, l) \in \overline{B_{\frac{1}{\alpha}}(0)}$ exactly whenever the respective point $\xi_j(\beta^+, l)$ of the reference orbit lies in the expanding region as well. In other words, a point $\hat{m} + j \in J^+(\hat{m})$ will be regular if and only if $j \in [0, \hat{l}^+]$ was regular before. This leads to a kind of self-similar structure in the sets of regular points,

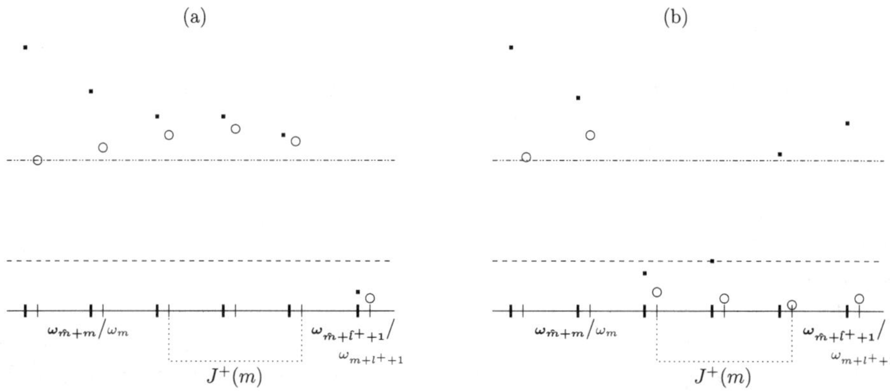

FIGURE 4.2. In the above diagram, $J(m)$ is located at the end of $[1, \hat{l}^+]$, such that $m + l^+ = \hat{l}^+$. At first, the new orbit $\xi_{\hat{m}+1}(\beta^+, l), \ldots, \xi_{\hat{m}+\hat{l}^++1}(\beta^+, l)$ will be pushed out of the expanding region (not shown). But at the end of the interval $[1, \hat{l}^+]$ the reference orbit $\xi_1(\beta^*, l), \ldots, \xi_{\hat{l}^++1}(\beta^*, l)$ leaves the expanding region for a few iterates. Thus, the new orbit may approach the reference orbit during this time and enter the expanding region again afterwards. Now we consider two different situations: In (a) we assume that the reference orbit spends all times $j \in J(m)$ outside of the expanding region. This is what we have to take into account if we do not define the points in $J^+(m)$ as regular, and consequently do not derive any information about them. Then the new orbit may still be close to the reference orbit until the very last step, and thus lie in the expanding region at the end. (b) On the other hand, if we can obtain information about the $j \in J^+(m)$ and thus define them as regular, then we know that the reference orbit stays in the expanding region at these times. Therefore the new orbit may enter the expanding region after time $\hat{m} + m$, but it will be pushed out again before the end of the interval $J(\hat{m})$ is reached.

which will express itself in relations of the following form:

$$R_N \cap J^+(\hat{m}) = \left(R_N \cap [1, \hat{l}^+]\right) + \hat{m} = R_{\hat{l}^+} + \hat{m} \tag{4.2}$$

In other words, the structure of the sets R_N after a close return, i.e. in the right part J^+ of an exceptional interval, is the same as their structure at the origin (see Figure 4.3).

What remains is to extend the construction not only forwards, but also backwards in time. As we have mentioned above, for some close return $\tilde{m}$ we will eventually have to choose $\tilde{l}^-$ larger than l. In this case, it is not sufficient anymore to have reference orbits starting on the ω_{-l}-fibre. However, we can still carry out the construction exactly up to $\tilde{m}$. Thus, if β^* is chosen such that $\xi_{\tilde{m}}(\beta^*, l) = 0$, then we will know that $\xi_{\tilde{m}-\tilde{l}^-}(\beta^*, l), \ldots, \xi_{\tilde{m}}(\beta^*, l)$ spends 'most' of the time in the expanding region. Therefore, we can use it as a reference orbit in order to show that $\xi_{-\tilde{l}^-}(\beta, \tilde{l}^-), \ldots,$

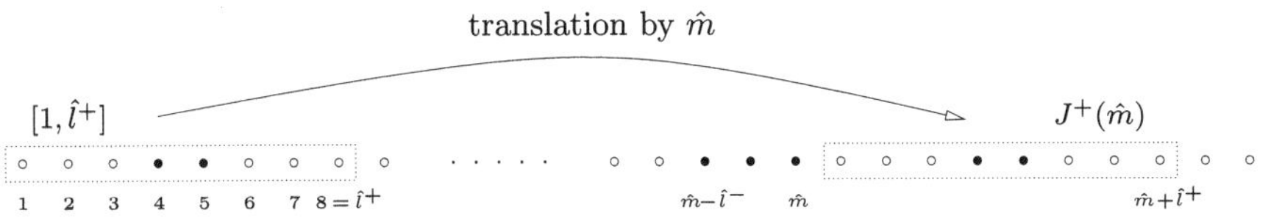

FIGURE 4.3. Recursive structure of the sets R_N. Regular points are shown in white, exceptional ones in black. The set $R_N \cap J^+(\hat{m})$ is a translate of the set $R_N \cap [1, \hat{l}^+]$.

$\xi_0(\beta, \tilde{l}^-)$ stays in the contracting region 'most' of the time, at least for parameters β which are close enough to β^*. (Recall that this orbit starts on the upper boundary line, i.e. $\xi_{-\tilde{l}^-}(\beta, \tilde{l}^-) = 3$ by definition.) It will then turn out that it suffices to consider such parameter values.

In this way, the construction will be extended backwards and we can then start to look at the forward part of the trajectories starting on the $\omega_{-\tilde{l}}$-fibre. Consequently, when we reach $\hat{m}$ again the backwards part of the trajectories is long enough to carry on beyond this point, again using the same comparison arguments as above. The only difference to Figure 4.1 will be that now the reference orbits only stay most and not all of the time in the expanding or contracting region, respectively. Nevertheless, this will still be sufficient to proceed more or less in the same way. Hence we can continue the construction, until we reach some even closer return. Then the trajectories have to be extended further in the backwards direction again and so on ... .

CHAPTER 5

Tools for the construction

In this section, we will provide the the necessary tools for the construction of the sink-source-orbits in Sections 6 and 7. As we have seen, there are mainly two things which have to be done: First, we need some statements about the comparison of orbits, namely one about expansion and one about contraction. These will be derived in Section 5.1. Secondly, we have to define the sets of admissible and regular times, which will be done in Sections 5.3 and 5.4 . However, before this we will have to introduce yet another collection of sets Ω_p ($p \in \mathbb{N}_0$) in Section 5.2. These sets Ω_p will be used as an approximation for the sets of *non-regular* times and will make it possible to control the frequency with which these can occur.

5.1. Comparing orbits

The two statements we aim at proving here are Lemma 5.2 and Lemma 5.6. They will allow to compare two different orbit-segments which (i) start on nearby fibres and (ii) result from systems T_{β_1}, T_{β_2} with parameters β_1, β_2 close together (compare Definition 4.4 and Remark 4.5). The reader should note that throughout this wubsection we only use assumptions (2.1),(2.2),(2.5)–(2.7) and the Lipschitz-continuity of g. In particular, we neither use the fact that g is non-negative, nor (2.11). Therefore, we will also be able to use the results for the case of symmetric forcing in Section 7. The diophantine condition on ω as well as (2.8) and (2.11) will not be needed until the next section. Before we start, we make one more assumption on the parameter α: We suppose that K is chosen as in Remark 4.5 and assume

$$\sqrt{\alpha} \geq 2K . \tag{5.1}$$

The following notation is tailored to our purpose of comparing two orbits:

DEFINITION 5.1. *Suppose T_β is defined as in Theorem 2.7 . If $\theta_1, \theta_2 \in \mathbb{T}^1$, $x_1^1, x_1^2 \in [-3,3]$ and $\beta_1, \beta_2 \in [0, \frac{3}{2}]$ are given, let*

$$x_n^1 := T_{\beta_1,\theta_1,n-1}(x_1^1) \quad , \quad x_n^2 := T_{\beta_2,\theta_2,n-1}(x_1^2) \tag{5.2}$$

and

$$\tau(n) := \#\{j \in [1,n] \mid x_j^1 \notin \overline{B_{\frac{1}{\alpha}}(0)}\} . \tag{5.3}$$

We start with a lemma about orbit-contraction. Essentially, the statement is that if two orbits spend most of the time in the contracting region above the line $\mathbb{T}^1 \times \{\gamma\}$, then their distance in the vertical direction gets contracted up to the magnitude of the error term:

LEMMA 5.2. *Suppose than conditions (2.5) and (2.7) hold and*

$$\mathrm{err}(\beta_1, \beta_2, \theta_1, \theta_2) \leq K \cdot \epsilon$$

for some $\epsilon > 0$. *Let further*

$$\eta(k,n) := \#\{j \in [k,n] \mid x_j^1 \text{ or } x_j^2 < \gamma\} \tag{5.4}$$

and assume that $\eta(j,n) \leq \frac{n+1-j}{10}$ $\forall j = 1,\ldots,n$ *and* $\alpha^{-\frac{n}{4}} \leq \epsilon$. *Then*

$$|x_{n+1}^1 - x_{n+1}^2| \;\leq\; \epsilon \cdot \left(6 + K \cdot \sum_{j=0}^{\infty} \alpha^{-\frac{1}{4}j}\right) . \tag{5.5}$$

A similar statement holds for $\tilde{\eta}(k,n) := \#\{j \in [k,n] \mid x_j^1 \text{ or } x_j^2 > -\gamma\}$.

PROOF. We prove the following statement by backwards induction on k: For all $k = 1,\ldots,n+1$ there holds

$$\begin{aligned} |x_{n+1}^1 - x_{n+1}^2| \;&\leq\; |x_k^1 - x_k^2| \cdot \alpha^{-\frac{1}{2}(n+1-k-5\eta(k,n))} \\ &\quad + K \cdot \epsilon \cdot \sum_{j=k+1}^{n+1} \alpha^{-\frac{1}{2}(n+1-j-5\eta(j,n))} \end{aligned} \tag{5.6}$$

The case $k = n+1$ is obvious. For the induction step, first suppose x_k^1 or $x_k^2 < \gamma$, such that $\eta(k,n) = \eta(k+1,n) + 1$. Then, by (2.5) we have

$$|x_{k+1}^1 - x_{k+1}^2| \leq |x_k^1 - x_k^2| \cdot \alpha^2 + K \cdot \epsilon \; ,$$

and by applying the statement for $k+1$ we get

$$\begin{aligned} &|x_{n+1}^1 - x_{n+1}^2| \;\leq \\ &\leq\; (|x_k^1 - x_k^2| \cdot \alpha^2 + K \cdot \epsilon) \cdot \alpha^{-\frac{1}{2}(n-k-5\eta(k+1,n))} + K \cdot \epsilon \sum_{j=k+2}^{n+1} \alpha^{-\frac{1}{2}(n+1-j-5\eta(j,n))} \\ &=\; |x_k^1 - x_k^2| \cdot \alpha^{-\frac{1}{2}(n+1-k-5\eta(k,n))} + K \cdot \epsilon \cdot \sum_{j=k+1}^{n+1} \alpha^{-\frac{1}{2}(n+1-j-5\eta(j,n))} \; . \end{aligned}$$

On the other hand, suppose $x_k^1, x_k^2 \geq \gamma$, such that $\eta(k,n) = \eta(k+1,n)$. In this case we can use (2.7) to obtain

$$|x_{k+1}^1 - x_{k+1}^2| \leq |x_k^1 - x_k^2| \cdot \alpha^{-\frac{1}{2}} + K \cdot \epsilon$$

and thus

$$\begin{aligned} |x_{n+1}^1 - x_{n+1}^2| \;&\leq\; (|x_k^1 - x_k^2| \cdot \alpha^{-\frac{1}{2}} + K \cdot \epsilon) \cdot \alpha^{-\frac{1}{2}(n-k-5\eta(k+1,n))} \\ &\quad + K \cdot \epsilon \cdot \sum_{j=k+2}^{n+1} \alpha^{-\frac{1}{2}(n+1-j-5\eta(j,n))} \\ &=\; |x_k^1 - x_k^2| \cdot \alpha^{\frac{1}{2}(n+1-k-5\eta(k,n))} \\ &\quad + K \cdot \epsilon \cdot \sum_{j=k+1}^{n+1} \alpha^{-\frac{1}{2}(n+1-j-5\eta(j,n))} \; . \end{aligned}$$

The statement of the lemma is now just an application of (5.6). Note that $|x_1^1 - x_1^2|$ is always bounded by 6.

□

The result about orbit-expansion we will need is a little bit more intricate. The problem is the following: We have one reference orbit, which spends most of the time well inside of the expanding region $\mathbb{T}^1 \times \overline{B_{\frac{2}{\alpha}}(0)}$. A second orbit starts a certain distance above, and we want to conclude that at some point it has to leave the expanding region while the first orbit remains inside at the same time. The following case is still quite simple:

LEMMA 5.3. *Suppose that conditions (2.2), (2.6) and (5.1) hold and further*

$$\operatorname{err}(\beta_1, \beta_2, \theta_1, \theta_2) \leq K \cdot \alpha^{-1}$$

and $x_1^2 \geq x_1^1 + \frac{1}{\alpha}$. *Then as long as* $\tau(n) = 0$ *there holds* $x_{n+1}^2 \geq x_{n+1}^1 + \frac{3}{\alpha}$. *Thus* $x_{n+1}^2 \geq \frac{2}{\alpha}$ *if* $x_{n+1}^1 \in \overline{B_{\frac{1}{\alpha}}(0)}$. *A similar statement holds if* $x_1^2 \leq x_1^1 - \frac{1}{\alpha}$.

PROOF. This follows from

$$x_{n+1}^2 \overset{(2.6)}{\geq} x_{n+1}^1 + 2\sqrt{\alpha} \cdot \tfrac{1}{\alpha} - K \cdot \alpha^{-1} \geq x_{n+1}^1 + \tfrac{1}{\alpha}(2\sqrt{\alpha} - K) \overset{(5.1)}{\geq} x_{n+1}^1 + \tfrac{1}{\sqrt{\alpha}}$$

as long as $x_n^2 - x_n^1 \geq \frac{1}{\alpha}$ and $x_n^1 \in \overline{B_{\frac{1}{\alpha}}(0)}$. Note that $\frac{1}{\sqrt{\alpha}} \geq \frac{3}{\alpha}$ by (2.2). $\square$

However, it is not always that easy, because we also need to address the case where the first orbit does not stay in the contracting region all but only 'most' of the times. This needs a little bit more care, and there are some natural limits: For example, x_j^1 must not spend to many iterates in the contracting region, even if these only make up a very small proportion of the length of the whole orbit segment. Otherwise the vertical distance between the two orbits may be contracted until it is of the same magnitude of the error term, and then the order of the orbits might get reversed. Another requirement is that x_j^1 does not leave the expanding region too often towards the end of the considered time interval. The reason for this was already demonstrated in Figure 4.2 .

In the end we aim at proving Lemma 5.6, which is the statement that will be used later on. However, in order to do so we need two intermediate lemmas first.

LEMMA 5.4. *Suppose that conditions (2.5), (2.6) and (5.1) hold and further*

$$\operatorname{err}(\beta_1, \beta_2, \theta_1, \theta_2) \leq K \cdot \epsilon$$

with $\epsilon \leq \alpha^{-q}$ *for some* $q \geq 1$ *and*

$$x_1^2 \geq x_1^1 + \frac{\epsilon}{2} \cdot \alpha^r \tag{5.7}$$

with $0 \leq r < q$. *Suppose further that for all* $j = 1, \ldots, n$ *there holds*

$$x_j^1 \in \overline{B_{\frac{1}{\alpha}}(0)} \Rightarrow x_j^2 \in \overline{B_{\frac{2}{\alpha}}(0)} \tag{5.8}$$

and

$$r + \tfrac{1}{2}(j - 5\tau(j)) \geq \tfrac{1}{2} . \tag{5.9}$$

Then

$$x_{n+1}^2 \geq x_{n+1}^1 + \frac{\epsilon}{2} \cdot \alpha^{r + \frac{1}{2}(n - 5\tau(n))} . \tag{5.10}$$

A similar statement holds if $x_1^2 \leq x_q^1 - \frac{\epsilon}{2} \cdot \alpha^r$.

Note that (5.9) is always guaranteed if either $\tau(n) \leq \max\{0, \frac{2r-1}{4}\}$ (as $5\tau(j) - j \leq 4\tau(j) \leq 4\tau(n)$), or if $\tau(j) \leq \frac{j}{8}\ \forall j = 1, \ldots, n$.

PROOF. We prove (5.10) by induction on n. The case $n = 0$ is obvious. For the induction step, we have to distinguish two cases:

Case 1: $\quad x_n^1 \in \overline{B_{\frac{1}{\alpha}}(0)}$, i.e. $\tau(n) = \tau(n-1)$

$$\begin{aligned} x_{n+1}^2 &\overset{(2.6)}{\geq} x_{n+1}^1 + 2\sqrt{\alpha} \cdot \frac{\epsilon}{2} \cdot \alpha^{r+\frac{1}{2}(n-1-5\tau(n-1))} - K \cdot \epsilon \\ &= x_{n+1}^1 + \epsilon \cdot \left(\alpha^{r+\frac{1}{2}(n-5\tau(n))} - K\right) \geq x_{n+1}^1 + \frac{\epsilon}{2} \cdot \alpha^{r+\frac{1}{2}(n-5\tau(n))} \end{aligned}$$

where we used $\alpha^{r+\frac{1}{2}(n-5\tau(n))} \geq \sqrt{\alpha} \geq 2K$ by (5.9) and (5.1) in the last step.

Case 2: $\quad x_n^1 \notin \overline{B_{\frac{1}{\alpha}}(0)}$, i.e. $\tau(n) = \tau(n-1) + 1$

$$\begin{aligned} x_{n+1}^2 &\overset{(2.5)}{\geq} x_{n+1}^1 + 2\alpha^{-2} \cdot \frac{\epsilon}{2} \cdot \alpha^{r+\frac{1}{2}(n-1-5\tau(n-1))} - K \cdot \epsilon \\ &= x_{n+1}^1 + \epsilon \cdot \left(\alpha^{r+\frac{1}{2}(n-5\tau(n))} - K\right) \geq x_{n+1}^1 + \frac{\epsilon}{2} \cdot \alpha^{r+\frac{1}{2}(n-5\tau(n))} \end{aligned}$$

where we used $\alpha^{r+\frac{1}{2}(n-5\tau(n))} \geq 2K$ again in the step to the last line.

□

LEMMA 5.5. *Suppose that conditions (2.2), (2.6) and (5.1) hold and*

$$\operatorname{err}(\beta_1, \beta_2, \theta_1, \theta_2) \leq K \cdot \alpha^{-q}$$

for some $q \geq 1$. *Further, assume that* $x_1^1, x_{n+1}^1 \in \overline{B_{\frac{1}{\alpha}}(0)}$, $x_1^2 \geq \frac{2}{\alpha}$ *and* $\tau(n) \leq \max\{0, \frac{2q-3}{4}\}$. *Then* $x_j^2 \geq x_j^1$ $\forall j = 1, \ldots, n$ *and there holds*

$$\underbrace{\#\left\{j \in [2, n+1] \mid x_j^1 \notin \overline{B_{\frac{1}{\alpha}}(0)} \text{ or } x_j^2 \in \overline{B_{\frac{2}{\alpha}}(0)}\right\}}_{=:\Upsilon} \leq 5\tau(n) .$$

A similar statement holds if $x_1^2 \leq -\frac{2}{\alpha}$.

PROOF. It suffices to obtain a suitable upper bound on $\#\tilde{\Upsilon}$ where

$$\tilde{\Upsilon} := \{j \in [2, n+1] \mid x_j^1 \in \overline{B_{\frac{1}{\alpha}}(0)} \text{ and } x_j^2 \in \overline{B_{\frac{2}{\alpha}}(0)}\} ,$$

as obviously $\#\Upsilon = \#\tilde{\Upsilon} + \tau(n+1) = \#\tilde{\Upsilon} + \tau(n)$. (Note that $\tau(n+1) = \tau(n)$ as $x_{n+1}^1 \in \overline{B_{\frac{1}{\alpha}}(0)}$ by assumption.) To that end, we can separately consider blocks $[k+1, l]$ where k, l are chosen such that

(i) $1 \leq k < l \leq n+1$
(ii) $x_j^1 \in \overline{B_{\frac{1}{\alpha}}(0)} \Rightarrow x_j^2 \in \overline{B_{\frac{2}{\alpha}}(0)}$ $\forall j \in [k+1, l]$
(iii) $x_k^1 \in \overline{B_{\frac{1}{\alpha}}(0)}$ and $x_k^2 \notin \overline{B_{\frac{2}{\alpha}}(0)}$
(iv) $x_l^1 \in \overline{B_{\frac{1}{\alpha}}(0)}$ and $x_l^2 \in \overline{B_{\frac{2}{\alpha}}(0)}$
(v) l is the maximal integer in $[k+1, n+1]$ with the above properties (ii) and (iv).

Note that $\tilde{\Upsilon}$ is contained in the disjoint union of all such blocks $[k+1, l]$.

We now want to apply Lemma 5.4, but starting with x_k^i instead of x_1^i $(i = 1, 2)$. Therefore, let $\tilde{\theta}_i = \theta_i + \omega_{k-1}$, $\tilde{x}_1^i = x_k^i$ and $\tilde{n} = l - k$ in Definition 5.1. Note that $\tilde{\tau}(\tilde{n}) = \tau(l-1) - \tau(k-1)$, but as we assumed that $x_k^1, x_l^1 \in \overline{B_{\frac{1}{\alpha}}(0)}$ in (iii) and

(iv) we equally have $\tilde{\tau}(\tilde{n}) = \tau(l) - \tau(k)$. As $x_k^2 \geq x_k^1 + \frac{1}{\alpha}$ by (iii), we can apply Lemma 5.4 with $\epsilon = \alpha^{-q}$ and $r = q-1$ to obtain

$$x_l^2 \;=\; \tilde{x}_{\tilde{n}+1}^2 \;\geq\; \tilde{x}_{\tilde{n}+1}^1 + \frac{\alpha^{-1}}{2} \cdot \alpha^{\frac{1}{2}(\tilde{n}-\tilde{\tau}(\tilde{n}))} \;=\; x_l^1 + \frac{\alpha^{-1}}{2} \cdot \alpha^{\frac{1}{2}(l-k-5(\tau(l)-\tau(k)))} \;.$$

As $|x_l^2 - x_l^1| \leq \frac{3}{\alpha} < \frac{1}{2\sqrt{\alpha}}$ by (iv) and (2.2), we must therefore have $l-k-5(\tau(l)-\tau(k)) \leq 0$ or equivalently $l-k \;\leq\; 5(\tau(l)-\tau(k))$. Thus

$$\#\left(\tilde{\Upsilon} \cap [k+1, l]\right) \;=\; l-k-(\tau(l)-\tau(k)) \;\leq\; 4(\tau(l)-\tau(k)) \;.$$

Summing over all such blocks $[k+1,l]$ we obtain $\#\tilde{\Upsilon} \leq 4\tau(n)$, and this completes the proof.

□

LEMMA 5.6. *Suppose that conditions (2.2), (2.5), (2.6) and (5.1) hold and*

$$\mathrm{err}(\beta_1, \beta_2, \theta_1, \theta_2) \leq K \cdot \epsilon$$

for some $\epsilon \in [\alpha^{-(q+1)}, \alpha^{-q})$, $q \geq 1$. *Further, assume that for some* $n \in \mathbb{N}$ *with* $x_{n+1}^1 \in \overline{B_{\frac{1}{\alpha}}(0)}$ *there holds* $\tau(n) \leq \max\{0, \frac{2q-3}{4}\}$ *and*

$$\tau(n) - \tau(j) \leq \frac{n-j}{6} \quad \forall j \in [1,n] \;. \tag{5.11}$$

Then

(a) $x_1^1 \in \overline{B_{\frac{1}{\alpha}}(0)}$ *but* $x_1^2 \geq \frac{2}{\alpha}$ *implies* $x_{n+1}^2 \geq \frac{2}{\alpha}$.

(b) *If* $n \geq 5q$ *and* $\tau(j) \leq \frac{j}{8}$ $\forall j = 1, \ldots, n$, *then* $x_1^2 \geq x_1^1 + \frac{\epsilon}{2}$ *implies* $x_{n+1}^2 \geq \frac{2}{\alpha}$.

Again, similar statements hold for the reverse inequalities.

PROOF.

(a) Note that $\tau(1) = 0$ as $x_1^1 \in \overline{B_{\frac{1}{\alpha}}(0)}$ by assumption. By Lemma 5.5 we have

$$\#\Upsilon \;\leq\; 5\tau(n) \;\overset{(5.11)}{\leq}\; \frac{5(n-1)}{6} \;\leq\; n-1 \;,$$

Thus there exists $j_0 \in [2, n+1]$ such that $x_{j_0}^1 \in \overline{B_{\frac{1}{\alpha}}(0)}$ but $x_{j_0}^2 \geq \frac{2}{\alpha}$.

If we shift the starting points in Definition 5.1 to $\tilde{\theta}_i := \theta_i + \omega_{j_0-1}$ and $\tilde{x}_1^i = x_{j_0}^i$ $(i=1,2)$ and denote the resulting sequences by $\tilde{x}_j^1, \tilde{x}_j^2$, then $\tilde{n} := n - j_0 + 1$ satisfies the same assumptions as before. As $\tilde{n} < n$ we can repeat this procedure until $\tilde{n} < 6$. But then $\tilde{\tau}(\tilde{n}) = 0$, such that $\tilde{x}_1^1 \in \overline{B_{\frac{1}{\alpha}}(0)}$ and $\tilde{x}_1^2 \geq \frac{2}{\alpha}$ implies $\tilde{x}_{\tilde{n}+1}^2 = x_{n+1}^2 \geq \frac{2}{\alpha}$ by Lemma 5.3, proving statement (a).

(b) We claim that there exists $j_1 \in [1, n+1]$ such that $x_{j_1}^1 \in \overline{B_{\frac{1}{\alpha}}(0)}$ but $x_{j_1}^2 \notin \overline{B_{\frac{2}{\alpha}}(0)}$.

Suppose there exists no such j_1 and let k be the largest integer in $[1,n]$ such that $x_{k+1}^1 \in \overline{B_{\frac{1}{\alpha}}(0)}$. As $x_j^1 \in \overline{B_{\frac{1}{\alpha}}(0)} \Rightarrow x_j^2 \in \overline{B_{\frac{2}{\alpha}}(0)}$ holds for all $j = 1, \ldots, k$, we can apply Lemma 5.4 with $r = 0$ to obtain

$$x_{k+1}^2 \;\geq\; x_{k+1}^1 + \frac{\epsilon}{2} \cdot \alpha^{\frac{1}{2}(k-5\tau(k))} \;\geq\; x_{k+1}^1 + \frac{1}{2}\alpha^{\frac{1}{2}(k-5\tau(k))-q-1} \;. \tag{5.12}$$

Now $\tau(n) = \tau(k+1)+n-k-1$ by definition of k. Further $\tau(k) = \tau(k+1)$, as $x^1_{k+1} \in \overline{B_{\frac{1}{\alpha}}(0)}$ by the choice of k. Therefore

$$\begin{aligned}
\frac{1}{2}(k - 5\tau(k)) \; &= \\
&= \; \frac{1}{2}(k+1-5\tau(k+1)) - \frac{1}{2} \; \geq \; \tfrac{1}{2}(n - 5\tau(n)) - \tfrac{1}{2} \\
&\geq \; \frac{1}{2}(5q - 5\cdot\frac{2q-2}{4}) - \frac{1}{2} \; = \; \frac{5}{2}q - \frac{5}{4}q + \frac{12}{4} - \frac{1}{2} \; > \; q \,.
\end{aligned}$$

Plugged into (5.12) this yields $x^2_{k+1} \geq x^1_{k+1} + \frac{1}{2}$, contradicting $x^1_{k+1} \in \overline{B_{\frac{1}{\alpha}}(0)}$ and $x^2_{k+1} \in \overline{B_{\frac{2}{\alpha}}(0)}$.

Thus we can choose j_1 with $x^1_{j_1} \in \overline{B_{\frac{1}{\alpha}}(0)}$ and $x^2_{j_1} \geq \frac{2}{\alpha}$ as claimed. Shifting the starting points as before we can now apply (a) to complete the proof.

□

5.2. Approximating sets

As mentioned in Section 4, for each close return $m \in \mathbb{N}$ with $d(\omega_m, 0) \leq \frac{3\gamma}{L_2}$ we will introduce an exceptional interval $J(m)$. However, before we can do so we first have to define some intermediate intervals $\Omega_p(m)$. These will contain the intervals $J(m)$, such that they can be used to obtain estimates on the 'density' of the union of exceptional intervals. As we need a certain amount of flexibility, we have to introduce a whole sequence of such approximating sets $(\Omega_p(m))_{p\in\mathbb{N}_0}$, which will be increasing in p.

The statements of this as well as the two following subsections do not involve the dynamics of T, they are only related to the underlying irrational rotation by ω. Therefore, the only assumptions which are used are the Diophantine condition (2.13) as well as (2.1) and (2.2).

DEFINITION 5.7. *(a) Let*

$$S_n(\alpha) \; := \; \begin{cases} \sum_{i=0}^{n-1} \alpha^{-i} & \text{if} \quad n \in \mathbb{N}\cup\{\infty\} \\ 1 & \text{if} \quad n \leq 0 \end{cases} .$$

(b) For $p \in \mathbb{N}_0 \cup \{\infty\}$ let $Q_p : \mathbb{Z} \to \mathbb{N}_0$ be defined by

$$Q_p(j) := \begin{cases} q \;\; \text{if} \;\; d(\omega_j, 0) \in \left[S_{p-q+1}(\alpha)\cdot\frac{\alpha^{-q}}{L_2}, S_{p-q+2}(\alpha)\cdot\frac{\alpha^{-(q-1)}}{L_2}\right) \;\; \text{for } q \geq 2 \\ 1 \;\; \text{if} \;\; d(\omega_j, 0) \in \left[S_p(\alpha)\cdot\frac{\alpha^{-1}}{L_2}, \frac{4\gamma}{L_2} + S_p(\alpha)\cdot\frac{\alpha^{-1}}{L_2}\cdot(1-\mathbf{1}_{\{0\}}(p))\right) \\ 0 \;\; \text{if} \;\; d(\omega_j, 0) \geq \frac{4\gamma}{L_2} + S_p(\alpha)\cdot\frac{\alpha^{-1}}{L_2}\cdot(1-\mathbf{1}_{\{0\}}(p)) \end{cases}$$

if $j \in \mathbb{Z}\setminus\{0\}$ and $Q_p(0) := 0$. Further let

$$p(j) \; := \; Q_0(j) \,.$$

(c) For fixed $u, v \in \mathbb{N}$ let $\tilde{u} := u+2$ and $\tilde{v} := v+2$. Then, for any $j \in \mathbb{Z}$ define

$$\Omega^-_p(j) := [j - \tilde{u}\cdot Q_p(j), j] \,, \quad \Omega^+_p(j) := [j+1, j+\tilde{v}\cdot Q_p(j)]$$

and

$$\Omega_p(j) := \Omega_p^-(j) \cup \Omega_p^+(j)$$

if $Q_p(j) > 0$*, with all sets being defined as empty if* $Q_p(j) = 0$*. Further let*

$$\Omega_p^{(\pm)} := \bigcup_{j\in\mathbb{Z}} \Omega_p^{(\pm)}(j) \qquad \text{and} \qquad \tilde{\Omega}_p^{(\pm)} := \bigcup_{\substack{j\in\mathbb{Z}\\ Q_p(j)\leq p}} \Omega_p^{(\pm)}(j)\ .$$

(d) *Finally, let*

$$\begin{aligned} \nu(q) &:= \min\{j\in\mathbb{N} \mid p(j)\geq q\} && \forall q\in\mathbb{N} \\ \tilde{\nu}(q) &:= \min\left\{j\in\mathbb{N} \mid d(\omega_j,0) < 3S_\infty(\alpha)\cdot \tfrac{\alpha^{-(q-1)}}{L_2}\right\} && \text{if } q\geq 2 \text{ and} \\ \tilde{\nu}(1) &:= \min\left\{j\in\mathbb{N} \mid d(\omega_j,0) < 3\left(\tfrac{4\gamma}{L_2} + S_\infty(\alpha)\cdot\tfrac{\alpha^{-1}}{L_2}\right)\right\}. \end{aligned}$$

REMARK 5.8. Suppose that (2.1) and (2.2) hold, such that we have $\sqrt{\alpha} \geq \frac{4}{\gamma} \geq 64$. As $S_\infty(\alpha) = \frac{\alpha}{\alpha-1}$, the following estimates can be deduced easily from this:

$$\alpha \geq S_\infty(\alpha)+1 \tag{5.1}$$

$$\gamma \geq \frac{S_\infty(\alpha)+1}{\alpha}\ . \tag{5.2}$$

REMARK 5.9. As in the prededing remark, we suppose that (2.1) and (2.2) hold.

(a) By definition, we have $Q_{p'}(j) \leq Q_p(j)\ \forall j\in\mathbb{Z}$ whenever $p'\leq p$. Further, there holds $Q_\infty(j)\leq p(j)+1\ \forall j\in\mathbb{N}$. For $p(j)\geq 1$ this follows from (5.1), which implies $\frac{S_\infty(\alpha)}{\alpha}\leq 1$. In the case $p(j)=0$ this is true by (5.2). Altogether, this yields

$$p(j) \leq Q_p(j) \leq Q_\infty(j) \leq p(j)+1 \quad \forall j\in\mathbb{Z},\ p\in\mathbb{N} \tag{5.3}$$

(b) As a direct consequence of (a) we have $\Omega_{p'}^{(\pm)}(j) \subseteq \Omega_p^{(\pm)}(j)\ \forall j\in\mathbb{N}$ whenever $p'\leq p$. The same holds for the sets $\Omega_p^{(\pm)}$ and $\tilde{\Omega}_p^{(\pm)}$.

The following two lemmas collect a few basic properties of the sets $\Omega_p^{(\pm)}$ and $\tilde{\Omega}_p^{(\pm)}$. The first one is a certain 'almost invariance' property under translations with m if ω_m is close to 0. This is closely related to the recursive structure of the sets R_N of regular points mentioned in the last section (see (4.2)), and explains why we had to introduce a whole family $(\Omega_p)_{p\in\mathbb{N}_0}$ of approximating sets.

Lemma 5.11 then contains the estimates which can be obtained from the diophantine condition. These will allow us to control the "density" the sets of $\Omega_\infty^{(\pm)}$ (and thus of the sets R_N defined later on) by making suitable assumptions on the parameters.

LEMMA 5.10. *Suppose that conditions (2.1) and (2.2) hold. Let* $p\geq 2$ *and suppose* $p(m)\geq p$ *and* $Q_{p-2}(k)\leq p-2$*. Then*

(a) $Q_{p-2}(k) \leq Q_{p-1}(k\pm m) \leq Q_{p-2}(k)+1$

(b) $\tilde{\Omega}_{p-2}^{(\pm)} \pm m \subseteq \tilde{\Omega}_{p-1}^{(\pm)}$. *Using* $\tilde{\Omega}_{-1} := \emptyset = \tilde{\Omega}_0$*, this also holds if* $p=1$.

PROOF.

(a) Let $q := Q_{p-2}(k)$, so that $p - q \geq 2$ by assumption. We first show that

(5.4) $$Q_{p-1}(k+m) \;\geq\; q\ .$$

To that end, first suppose $q \geq 2$, such that $d(\omega_k, 0) < S_{p-q}(\alpha) \cdot \frac{\alpha^{-(q-1)}}{L_2}$. Then

$$d(\omega_{k\pm m}, 0) \;\leq\; d(\omega_k, 0) + d(\omega_m, 0) \;<\; S_{p-q}(\alpha) \cdot \frac{\alpha^{-(q-1)}}{L_2} + \frac{\alpha^{-(p-1)}}{L_2}$$
$$= \;\left(S_{p-q}(\alpha) + \alpha^{-(p-q)}\right) \cdot \frac{\alpha^{-(q-1)}}{L_2} \;=\; S_{p-q+1}(\alpha) \cdot \frac{\alpha^{-(q-1)}}{L_2}\ .$$

This proves (5.4) in case $q \geq 1$. The case $q = 1$ is treated similarly, if $q = 0$ there is nothing to show.

It remains to prove that

(5.5) $$Q_{p-1}(k+m) \;\leq\; q+1\ .$$

This time, first assume $q \geq 1$, such that $d(\omega_k, 0) \geq S_{p-q-1}(\alpha) \cdot \frac{\alpha^{-q}}{L_2}$. Then

$$d(\omega_{k\pm m}, 0) \;\geq\; d(\omega_k, 0) - d(\omega_m, 0) \;\geq\; S_{p-q-1}(\alpha) \cdot \frac{\alpha^{-q}}{L_2} - \frac{\alpha^{-(p-1)}}{L_2}$$
$$= \;\underbrace{\left(\alpha \cdot S_{p-q-1}(\alpha) - \alpha^{-(p-q-2)}\right)}_{\geq\, \alpha - 1 \,\geq\, S_\infty(\alpha) \text{ by } (5.1)} \cdot \frac{\alpha^{-(q+1)}}{L_2} \;\geq\; S_{p-q-1}(\alpha) \cdot \frac{\alpha^{-(q+1)}}{L_2}\ .$$

This implies (5.5). Again, the case $q = 0$ is treated similarly, using (5.2) instead of (5.1).

(b) Now suppose $j \in \tilde{\Omega}^{(\pm)}_{p-2}$. Then $\exists k \in \mathbb{Z}$ such that $Q_{p-2}(k) \leq p-2$ and $j \in \Omega^{(\pm)}_{p-2}(k)$. As $Q_{p-1}(k \pm m) \geq Q_{p-2}(k)$ by (a), this implies $j \pm m \in \Omega^{(\pm)}_{p-1}(k \pm m)$, and as $Q_{p-1}(k+m) \leq Q_{p-2}(k) + 1 \leq p-1$ this set is contained in $\tilde{\Omega}^{(\pm)}_{p-1}$.

□

LEMMA 5.11. *Let $u, v \in \mathbb{N}$ be fixed and suppose ω satisfies the diophantine condition (2.13). Then there exist functions $h, H : \mathbb{R}^2_+ \to \mathbb{R}_+$ with $h(\gamma, \alpha) \nearrow \infty$ and $H(\gamma, \alpha) \searrow 0$ as $(\gamma + \alpha^{-1}) \searrow 0$, such that*

(a)

$$\nu(q) \;\geq\; \tilde{\nu}(q) \;\geq\; h(\gamma, \alpha) \cdot (q+2) \cdot w \quad \forall q \in \mathbb{N}$$

where $w := \tilde{u} + \tilde{v} + 1 = u + v + 5$.

(b)

$$\#([1, N] \cap \Omega_\infty) \;\leq\; H(\gamma, \alpha) \cdot N \;\; \textit{and} \;\; \#([-N, -1] \cap \Omega_\infty) \;\leq\; H(\gamma, \alpha) \cdot N \quad \forall N \in \mathbb{N}.$$

PROOF.

(a) The diophantine condition implies that $c \cdot \tilde{\nu}(q)^{-d} \leq 2S_\infty(\alpha) \cdot \frac{\alpha^{-(q-1)}}{L_2}$ (if $q \geq 2$). Thus, a simple calculation yields

$$\frac{\tilde{\nu}(q)}{w \cdot (q+2)} \;\geq\; \left(\frac{c \cdot L_2}{2S_\infty(\alpha)}\right)^{\frac{1}{d}} \cdot \frac{\alpha^{\frac{q-1}{d}}}{w \cdot (q+2)}$$

and the right hand side converges to ∞ uniformly in q as $\alpha \to \infty$. Similarly,

$$\frac{\tilde{\nu}(1)}{3w} \geq \frac{1}{3w} \cdot \left(\frac{(c \cdot L_2)}{(8\gamma + 2S_\infty(\alpha) \cdot \alpha^{-1})} \right)^{\frac{1}{d}}$$

and again the right hand side converges to ∞ as $\gamma + \alpha^{-1} \searrow 0$. Thus we can define the minimum of both estimates as $h(\gamma, \alpha)$.

(b) As we have seen in (a) we have $\tilde{\nu}(q) \geq \left(\frac{c \cdot L_2}{2S_\infty(\alpha)}\right)^{\frac{1}{d}} \cdot \alpha^{\frac{q-1}{d}}$ if $q \geq 2$. Now $[1, N] \cap \Omega_\infty(j) = \emptyset$ if $j > N + w \cdot Q_\infty(j)$. Therefore

$$\begin{aligned} &\tfrac{1}{N} \cdot \#([1, N] \cap \Omega_\infty) \\ \leq\ & \frac{1}{N} \sum_{q=1}^{\infty} q \cdot w \cdot \#\{1 \leq j \leq N + q \cdot w \mid Q_\infty(j) = q\} \\ \leq\ & \frac{1}{N} \left(\frac{N + w}{\tilde{\nu}(1)} \cdot w + \sum_{q=2}^{\infty} q \cdot w \cdot \frac{N + q \cdot w}{\tilde{\nu}(q)} \right) \\ \leq\ & \frac{w + \frac{w^2}{N}}{\tilde{\nu}(1)} + \sum_{q=2}^{\infty} \frac{q \cdot w + q^2 \cdot \frac{w^2}{N}}{\left(\frac{c \cdot L_2}{2S_\infty(\alpha)}\right)^{\frac{1}{d}} \cdot \alpha^{\frac{q-1}{d}}} . \end{aligned}$$

The right hand side converges to 0 uniformly in N as $\gamma + \alpha^{-1} \to 0$ and we can use it as the definition of $H(\gamma, \alpha)$.

□

5.3. Exceptional intervals and admissible times

In order to decide whether a time $N \in \mathbb{N}$ is admissible, in the sense of Section 4.3, we will first have to introduce exceptional intervals $J(m)$ corresponding to close returns $m \in \mathbb{N}$ with $d(\omega_m, 0) \leq \frac{3\gamma}{L_2}$. For the sets Ω_p defined above, two different intervals $\Omega_p(m)$ and $\Omega_p(n)$ $(m \neq n)$ can overlap, without one of them being contained in the other. This is something we want to exclude for the exceptional intervals, and we can do so by carefully choosing their lengths. To this end, we have to introduce two more assumptions on the parameters:

We let h and H be as in Lemma 5.11 and suppose that γ and α are chosen such that $h(\gamma, \alpha) \geq 1$ and $H(\gamma, \alpha) \leq \frac{1}{12w}$. In other words, we will assume that for all $q, N \in \mathbb{N}$ there holds

$$\tilde{\nu}(q) \geq (q + 2) \cdot w , \tag{5.1}$$

$$\#([-N, -1] \cap \Omega_\infty) \leq \frac{N}{12w} \quad \textit{and} \quad \#([1, N] \cap \Omega_\infty) \leq \frac{N}{12w} . \tag{5.2}$$

REMARK 5.12. Suppose that (2.1), (2.2) and (5.1) hold. Assumption (5.1) ensures that on the one hand the sets $\Omega_\infty(j)$ never contain the origin (and are, in fact, a certain distance away from it), and on the other hand two such sets of approximately equal size do not interfere with each other. This will be very convenient later on. To be more precise:

(a) There holds

$$-2, -1, 0, 1, 2 \notin \Omega_\infty . \tag{5.3}$$

(b) If $Q_\infty(j) \geq q$ for some $j \in \mathbb{Z}$ then

$$[-\tilde{u} \cdot (q+2), \tilde{v} \cdot (q+2)] \cap \Omega_\infty(j) = \emptyset \; . \tag{5.4}$$

(c) Let $m, n \in \mathbb{Z}$, $m \neq n$. Then $\Omega_\infty(m) \cap \Omega_\infty(n) = \emptyset$ whenever $|Q_\infty(m) - Q_\infty(n)| \leq 2$ or $|Q_p(m) - Q_p(n)| \leq 1$ for some $p \in \mathbb{N}_0$.

PROOF. (a) and (b) follow immediately from (5.1) and the definition of the sets $\Omega_\infty(j)$. In order to prove (c), let $q := \min\{Q_\infty(m), Q_\infty(n)\}$. Then necessarily $d(\omega_{m-n}, 0) = d(\omega_m, \omega_n) < 2S_\infty(\alpha) \cdot \frac{\alpha^{-(q-1)}}{L_2}$ and thus $|m-n| \geq \tilde{\nu}(q) \geq (q+2) \cdot w$ by (5.1). On the other hand both $Q_\infty(m)$ and $Q_\infty(n)$ are at most $q+2$, and thus the definition of the $\Omega_\infty(j)$ implies the disjointness of the two sets. Finally, note that $|Q_p(m) - Q_p(n)| \leq 1$ implies $|Q_\infty(m) - Q_\infty(n)| \leq 2$ by (5.3).

□

REMARK 5.13. Suppose that (2.1), (2.2), (5.1) and (5.2) hold. (5.2) ensures that the "density" of the set Ω_∞ is small enough, and this will be very important for the construction later on. On the other hand, it also enables us now to choose suitable lengths for the exceptional intervals $J(m)$:

We have $\#([-\tilde{u} \cdot q, -1] \cap \Omega_\infty) \leq \frac{q}{12}$. This implies that we can find at least two consecutive integers outside of Ω_∞ in the interval $[-\tilde{u} \cdot q, -u \cdot q]$. In other words, for all $q \in \mathbb{N}$ there exists $l_q^- \in \mathbb{N}$ such that

$$u \cdot q \;\leq\; l_q^- \;<\; \tilde{u} \cdot q \quad \text{and} \quad -l_q^-,\; -l_q^- - 1 \notin \Omega_\infty \; . \tag{5.5}$$

Similarly, there exists $l_q^+ \in \mathbb{N}$, such that

$$v \cdot q \;\leq\; l_q^+ \;<\; \tilde{v} \cdot q \quad \text{and} \quad l_q^+,\; l_q^+ + 1 \notin \Omega_\infty \; . \tag{5.6}$$

In addition, we can assume that $l_p^\pm \geq l_q^\pm$ whenever $p \geq q$. (If $l_q^+, l_q^+ + 1$ are both contained in $[v \cdot (q+1), \tilde{v} \cdot (q+1)]$, then we can just take $l_{q+1}^+ = l_q^+$. Otherwise, we find a suitable $l_{q+1}^+ > l_q^+$ in this interval.) Note also that (5.5),(5.6) and (5.1) together imply that

$$\min\{u, v\} \cdot q \;\leq\; l_q^\pm \;<\; \tilde{\nu}(\max\{1, q-2\}) \;\leq\; \nu(\max\{1, q-2\}) \; . \tag{5.7}$$

Now we are able to define the exceptional intervals:

DEFINITION 5.14 (Exceptional intervals). *Suppose that (2.1), (2.2), (5.1) and (5.2) hold. Then for any $q \in \mathbb{N}$, choose $l_q^\pm$ as in Remark 5.13 and define, for any $m \in \mathbb{N}$ with $p(m) \geq 0$,*

$$\lambda^-(m) \;:=\; m - l_{p(m)}^- \quad , \quad \lambda^+(m) \;:=\; m + l_{p(m)}^+$$

$$J^-(m) \;:=\; [\lambda^-(m), m] \quad , \quad J^+(m) \;:=\; [m+1, \lambda^+(m)]$$

and

$$J(m) \;:=\; J^-(m) \cup J^+(m) \; .$$

If $p(m) = -1$, then $J^{(\pm)} := \emptyset$. Further, let

$$A_N \;:=\; [1, N] \setminus \bigcup_{1 \leq m < N} J(m) \quad \textit{and} \quad \Lambda_N \;:=\; [1, N] \setminus A_N$$

From now on, we will use conditions (2.1), (2.2), (5.1) and (5.2) as standing assumptions in the remainder of this subsection, as well as in Subsection 5.4 (since all the statements concern the preceding definition, directly or indirectly).

REMARK 5.15. (a) As we have mentioned before, the exceptional intervals are contained in the approximating sets. To be more precise, for each $m \in \mathbb{N}$ with $p(m) \geq 0$ there holds

$$(5.8)\qquad \begin{aligned} J(m) &\subset [\lambda^-(m) - 1, \lambda^+(m) + 1] \\ &\subseteq \Omega_0(m) \subseteq \Omega_p(m) \subseteq \Omega_\infty(m) , \end{aligned}$$

where $p \in \mathbb{N}$ is arbitrary. This follows from the choice of the $l_q^\pm$ in Remark 5.13 together with the definition of the intervals $\Omega_p(m)$. As a consequence, we have that

$$(5.9)\qquad \Lambda_N \subseteq \Omega_0 \subseteq \Omega_p \subseteq \Omega_\infty \quad \forall N, p \in \mathbb{N} .$$

(b) Further, suppose that $m \neq n$ and $|Q_\infty(m) - Q_\infty(n)| \leq 2$ or $|Q_p(m) - Q_p(n)| \leq 1$ for some $p \in \mathbb{N}_0$. Then (a) together with Remark 5.12(c) implies that

$$(5.10)\qquad \begin{aligned} J(m) \cap J(n) &= \emptyset = \\ &= [\lambda^-(m) - 1, \lambda^+(m) + 1] \cap [\lambda^-(n) - 1, \lambda^+(n) + 1] . \end{aligned}$$

In particular this is true if $|p(m) - p(n)| \leq 1$ (recall that $p(j) = Q_0(j)$).

(c) The sets A_N were defined as subsets of $[1, N]$, and it will turn out that they contain a very large proportion of the points from that interval. This could lead to this impression they form an increasing sequence of sets, but this is not true. For example, suppose that N itself is a close return, such that $p(N) \geq 1$. In this case N may still be contained in A_N, as the exceptional interval $J(N)$ is not taken into account in the definition of this set, but surely $N \notin A_{N+1}$. Thus, whenever we reach a close return, there may be a sudden decrease in the sets A_N in the next step. In general, we only have the two relations

$$(5.11)\qquad A_{N_2} \setminus A_{N_1} \subseteq [N_1 + 1, N_2] \quad \text{and}$$

and

$$(5.12)\qquad A_{N_2} \cap [1, N_1] \subseteq A_{N_1} .$$

where $N_1 \leq N_2$. However, the fluctuations and sudden decreases will only take place at the end of the interval $[1, N]$, and the starting sequence of the sets A_N will stabilize at some point: Suppose $N_0 \leq N_1 \leq N_2$ and $N_0 \in A_{N_2}$. Then

$$(5.13)\qquad A_{N_1} \cap [1, N_0] = A_{N_2} \cap [1, N_0] = A_{N_0} .$$

This simply follows from the fact that when N_0 is contained in A_{N_2} no exceptional interval $J(m)$ with $m \in [N_0 + 1, N_2 - 1]$ can reach into $[1, N_0]$, as it would then have to contain N_0. Thus $A_{N_1} \cap [1, N_0] = A_{N_2} \cap [1, N_0] = [1, N_0] \setminus \bigcup_{1 \leq m < N_0} J(m)$.

Note that (5.13) is always true whenever N_0 is not contained in any exceptional interval, i.e. $N_0 \notin \bigcup_{m \in \mathbb{N}} J(m) \subseteq \Omega_0 \subseteq \Omega_\infty$. In this case we have

$$(5.14)\qquad A_N \cap [1, N_0] = A_{N_0} \quad \forall N \geq N_0 .$$

In particular, as $l_q^+ \notin \Omega_\infty$, this implies $[1, l_q^+] \cap A_N = A_{l_q^+}\ \forall N \geq l_q^+$.

(d) Note also that it is not always true that $\Lambda_N = \bigcup_{1\leq m<N} J(m)$, as one of the exceptional intervals might extend beyond N, whereas Λ_N was defined as a subset of $[1, N]$. However, as we will see this relation holds as soon as we restrict to *'admissible'* times (see below).

The sets A_N will serve three different aims: First of all, they will play an important role in the construction of the sink-source-orbits themselves. Secondly, they will also be intermediates for the definition of the sets R_N of regular points. And finally, we will now use them to define admissible times:

DEFINITION 5.16 (Admissible times). *A time $N \in \mathbb{N}$ is called* ***admissible*** *if $N \in A_N$ (which is equivalent to $N \notin \Lambda_N$). The set $\{N \in \mathbb{N} \mid N \text{ is admissible }\}$ will be denoted by A.*

REMARK 5.17. (a) Any $N \in \mathbb{N}\backslash\Omega_0$ is admissible (see Remark 5.15(a)). In particular, l_q^+ and $l_q^+ + 1$ are admissible for any $q \geq 1$.

(b) As we mentioned above, for any admissible time N there holds

$$\Lambda_N = \bigcup_{1\leq m<N} J(m) , \tag{5.15}$$

as $N \in A_N$ ensures that none of the exceptional intervals $J(m)$ with $m < N$ extends further than $N-1$.

(c) For any $N_1 \in \mathbb{N}$, all times $N_0 \in A_{N_1}$ are admissible. This is a direct consequence of the fact that $A_{N_1} \cap [1, N_0] \subseteq A_{N_0}$ (see (5.12)). However, as already mentioned there might also be further admissible times contained in $[1, N_1] \setminus A_{N_1} = \Lambda_{N_1}$.

(d) Note that $A = \bigcup_{N\in\mathbb{N}} A_N$. The inclusion $\subseteq$ follows directly from the definition, whereas $\supseteq$ is a consequence of (c).

Now we can also verify the property of the exceptional intervals which was mentioned at the beginning of this section: Whenever two such intervals $J(m)$ and $J(n)$ intersect, one of them is contained in the other. We do not prove this statement in full, but rather concentrate on 'maximal' intervals, as this will be sufficient for our purposes.

LEMMA 5.18. *Let $N \in \mathbb{N}$ be admissible and suppose J is a non-empty maximal interval in $\Lambda_N = [1, N] \setminus A_N$. Then there exists a unique $m \in J$ with $p(m) = \max_{j\in J} p(j)$, and there holds $J = J(m)$. Furthermore, $p(j) < p - 1 \ \forall j \in J \setminus \{m\}$.*

PROOF. Let $p := \max_{j\in J} p(j)$ and $m \in J$ with $p(m) = p$. Obviously there holds $J(m) \subseteq J$. By definition, there cannot be any $j \in J \supseteq J(m)$ with $p(j) > p$. Therefore, as Remark 5.15(b) implies that $|p(j) - p| > 1 \ \forall j \in J(m) \setminus \{m\}$, there holds $p(j) < p-1 \ \forall j \in J(m) \setminus \{m\}$. Thus, it suffices to prove that $J = J(m)$. This will in turn follow if we can show that $\lambda^-(m) - 1$ and $\lambda^+(m) + 1$ are not contained in Λ_N, because then $J(m)$ is a maximal interval in Λ_N itself and must therefore be equal to J. We will only treat the case of $\lambda^-(m) - 1$, the other one is similar. In order to show that $\lambda^-(m) - 1$ is not contained in $J(k)$ for any $k = 1, \ldots, N$ we distinguish three different cases, according to the value of $Q_{p-2}(k)$:

First suppose $Q_{p-2}(k) > p + 1$. Then $p(k) > p$ by (5.3). If $\lambda^-(m) - 1 \in J(k)$, then $J(k) \cup J(m)$ is an interval and therefore $k \in J(k) \subseteq J$. But this contradicts the definition of p.

If $Q_{p-2}(k) \in \{p-1, p, p+1\}$, then $|Q_\infty(k) - Q_\infty(m)| \leq 2$ (again (5.3)) and therefore $\lambda^-(m) - 1 \notin \Lambda(k)$ by Remark 5.15(b).

This only leaves the possibility $Q_{p-2}(k) \leq p-2$. But in this case $\lambda^-(m) - 1 \in \Lambda(k)$ implies $\lambda^-(m) - 1 \in \Omega_{p-2}(k) \subseteq \tilde{\Omega}_{p-2}$ (see Remark 5.15(a)). As $p(m) = p$ we can apply Lemma 5.10(b) to obtain that $\lambda^-(m) - 1 - m = -l_p^- - 1 \in \tilde{\Omega}_{p-1}$, contradicting $-l_p^- - 1 \notin \Omega_\infty$ (by the choice of the $l_q^\pm$ in Remark 5.13).

As mentioned, the same arguments apply to $\lambda^+(m) + 1$, which completes the proof.

□

This naturally leads to the following

DEFINITION 5.19. *If N is admissible and $A_N = \{a_1, \ldots, a_n\}$ with $1 = a_1 < \ldots < a_n = N$, let*

$$\mathcal{J}_N := \{[a_k + 1, a_{k+1} - 1] \mid k = 1, \ldots, n-1\} \setminus \{\emptyset\}$$

be the family of all maximal intervals in $\Lambda_N = [1, N] \setminus A_N$ and $\mathcal{J} := \bigcup_{N \in \mathbb{N}} \mathcal{J}_N$. For any $J \in \mathcal{J}$ let $p_J := \max_{j \in J} p(j)$ and define m_J as the unique $m \in J$ with $p(m) = p_J$. m_J will be called the **central point** *of the interval J.*

Further, let $J^- := J^-(m_J)$ and $J^+ := J^+(m_J)$ (note that $J = J(m_J)$ by Lemma 5.18).

Note that not for every $n \in \mathbb{N}$ with $p(n) > 0$ the interval $J(n)$ is contained in $\mathcal{J}$. In fact, this will be wrong whenever $J(n) \subseteq J^+(m)$ for some $m < n$.

Among some other facts, the following lemma states that central points are always admissible. In the light of the discussion in Section 4.3, it is not surprising that this will turn out to be crucial for the construction.

LEMMA 5.20. *(a) Let $J \in \mathcal{J}$. Then $\lambda^-(m_J) - 1 \in A_{m_J}$, $\lambda^-(m_J) \in A_{m_J}$ and $m_J \in A_{m_J}$. In particular, $\lambda^-(m_J) - 1, \lambda^-(m_J)$ and m_J are admissible. Further, there holds*

$$p(j) \leq Q_\infty(j) \leq \max\{0, p_J - 2\} \quad \forall j \in J \setminus \{m_J\}\,. \tag{5.16}$$

(b) More generally, if $J \in \mathcal{J}$ and $q \leq p_J$, then $m_J - l_q^- - 1$, $m_J - l_q^-$, $m_J \in A_{m_J}$. In particular, $m_J - l_q^- - 1$, $m_J - l_q^-$ and m_J are admissible. Further there holds

$$p(j) \leq Q_\infty(j) \leq \max\{0, q-2\} \quad \forall j \in [m_J - l_q^-, m_J + l_q^+] \setminus \{m_J\}\,. \tag{5.17}$$

(c) If $J \in \mathcal{J}$, then $\lambda^+(m_J) + 1$ is admissible.

(d) For all $q \in \mathbb{N}$ there holds $\nu(q) - l_q^- - 1$, $\nu(q) - l_q^-$, $\nu(q) \in A_{\nu(q)}$ and

$$Q_\infty(j) \leq \max\{0, q-2\} \quad \forall j \in [\nu(q) - l_q^-, \nu(q) + l_q^+] \setminus \{\nu(q)\}\,.$$

In particular $\nu(q) - l_q^- - 1, \nu(q) - l_q^-$ and $\nu(q)$ are admissible.

PROOF.

(a) This is a special case of (b), which we prove below.

(b) Let $m := m_J$ and $j \neq m$. Suppose $Q_\infty(j) \geq q - 2$. Then

$$d(\omega_{m-j}, 0) = d(\omega_m, \omega_j) \leq 2S_\infty(\alpha) \cdot \frac{\alpha^{-(q-3)}}{L_2}\,.$$

Therefore $|m-j| \geq \tilde{\nu}(q-2) > l_q^{\pm}$ by (5.7), which implies that $j \notin [m-l_q^-, m+l_q^+]$. This proves (5.17).

As $J \in \mathcal{J}$, there exists some $N > m$ such that J is a maximal interval in Λ_N and consequently $\lambda^-(m)-1$ is contained in A_N (in particular $p(\lambda^-(m)-1)=0$). Hence, for any $n < \lambda^-(m)-1$ the interval $J(n)$ lies strictly to the left of $\lambda^-(m)-1$ and can therefore not intersect J. Thus, in order to show that $m-l_q^- - 1, l_q^-, m \in A_m$, it suffices to show that none of these points is contained in $U := \bigcup_{n\in[\lambda^-(m),m-1]} J(n)$. However, by (5.8) and (5.17) there holds $U \subseteq \tilde{\Omega}_{q-2}$. As $p(m) \geq q$ by assumption, Lemma 5.10(b) implies $U - m \subseteq \tilde{\Omega}_{q-1} \subseteq \Omega_\infty$ and the statement follows from $-l_q^- - 1, l_q^-, 0 \notin \Omega_\infty$.

Finally, note that $m - l_q^- \in A_m$ implies $m - l_q^- \in A_{m-l_q^-}$ by (5.12), similarly for $m - l_q^- - 1$, such that these points are both admissible.

(c) As m is admissible, $\lambda^+(m)+1$ cannot be contained in $J(n)$ for any $n < m$ (as all of these intervals must be contained in $[1, m-1]$). Thus, it suffices to show that $\lambda^+(m)+1$ is not contained in $\tilde{U} := \bigcup_{n\in[m+1,\lambda^+(m)]} J(n)$. But this set is again contained in $\tilde{\Omega}_{p_J-2}$ by (5.17). Therefore $\tilde{U} - m \subseteq \Omega_\infty$ by Lemma 5.10(b), and the statement follows from $l_{p_J}^+ + 1 \notin \Omega_\infty$.

(d) We show that $\nu(q)$ is admissible. Lemma 5.21 below then implies that $\nu(q)$ is a central point, and we can therefore apply (b) in order to prove (d).

Suppose $n < \nu(q)$. We have to show that $\nu(q) \notin \Omega_\infty(n) \supseteq J(n)$. In order to see this, note that $p(j) < q$ by definition of $\nu(q)$. Thus $d(\omega_{\nu(q)-n}, 0) = d(\omega_{\nu(q)}, \omega_n) \geq \tilde{\nu}(q) \geq (q+2)\cdot w$ by (5.1), and consequently $\nu(q) \notin \Omega_\infty(n)$. As $n < \nu(q)$ was arbitrary, this implies $\nu(q) \in A_{\nu(q)}$, such that $\nu(q)$ is admissible.

□

For Part (a) of the preceding lemma, the inverse is true as well:

LEMMA 5.21. *Suppose $m \in \mathbb{N}$ is admissible and $p(m) > 0$. Then $J(m) \in \mathcal{J}_{\lambda^+(m)+1} \subseteq \mathcal{J}$ and $\lambda^-(m)-1, \lambda^-(m)$ and $\lambda^+(m)+1$ are admissible.*

PROOF. We start by proving that $\lambda^+(m)+1$ is admissible, i.e. contained in $A_{\lambda^+(m)+1}$. First of all, the fact that m is admissible ensures that none of the intervals $J(n)$ with $n < m$ intersects $[m+1, \lambda^+(m)+1]$. Therefore, none of these intervals can contain $\lambda^+(m)+1$, and for $J(m)$ the same is true by definition. Now suppose $n \in [m+1, \lambda^+(m)]$. Then, similar as in the proof of Lemma 5.20(b) we obtain $p(n) \leq p(m)-2$ and therefore $J(n) \subseteq \tilde{\Omega}_{p(m)-2}$. Thus $J(n)-m$ is contained in $\tilde{\Omega}_{p(m)-1} \subseteq \Omega_\infty$ by Lemma 5.10(b) and can therefore not contain $l_{p(m)}^+ + 1 \notin \Omega_\infty$. Thus $\lambda^+(m)+1 = m + l_{p(m)}^+ + 1$ is admissible.

By Lemma 5.18, for any maximal interval $J = J(n) \in \mathcal{J}_{\lambda^+(m)+1}$ that intersects $J(m)$ there holds either $J(n) = J(m)$, such that $n = m$, or $J(m) \subsetneq J(n)$. However, the second case cannot occur if $n < m$ (as m is admissible), and for $n \in [m+1, \lambda^+(m)]$ it is ruled out as we have just argued that $p(n) < p(m)$ for such n. This proves $J(m) \in \mathcal{J}_{\lambda^+(m)+1}$.

Finally, we can apply Lemma 5.20(a) to $J = J(m)$, which yields that $\lambda^-(m)-1$ and $\lambda^-(m)$ are admissible as well.

□

5.4. Regular times

Now we can turn to defining the sets of regular points $R_N \subseteq [1, N]$. The sets A_N already contain all points outside of the exceptional intervals $J(m)$ ($m \in [1, N-1]$). As described in Section 4, we have to add certain points from the right parts $J^+(m)$ of these intervals. In order to do so, for each $J \in \mathcal{J}_N$ we will define a set $R(J) \subseteq J^+$ and then let $R_N = A_N \cup \bigcup_{J \in \mathcal{J}_N} R(J)$. Both R_N and $R(J)$ will be defined by induction on p. To be more precise, in the p-th step of the induction we first define $R(J)$ for all $J \in \mathcal{J}$ with $p_J \leq p-1$, and then R_N for all admissible times $N \leq \nu(p)$.

As in the preceding one, conditions (2.1), (2.2), (5.1) and (5.2) will be used as standing assumptions in this subsection (since all of the statements in this subsection directly or indirectly depend on Definition 5.14).

DEFINITION 5.22 (Regular times). *As mentioned, we proceed by induction on p. Note that the inclusions $R_N \subseteq [1, N]$ and $R(J) \subseteq J^+$ are preserved in every step of the induction.*

$\underline{p = 1:}$ *In order to start the induction let*

$$R_N := [1, N]$$

for any $N \leq \nu(1)$. Note that by definition there is no $J \in \mathcal{J}$ with $p_J = 0$.

$\underline{p \to p+1:}$ *Suppose $R(J)$ has been defined for all $J \in \mathcal{J}$ with $p_J \leq p - 1$ and R_N has been defined for all admissible times $N \leq \nu(p)$. In particular, this means that $R_{l_p^+}$ has defined already.*[1] *Then, for all $J \in \mathcal{J}$ with $p_J = p$ let*

$$R(J) = R_{l_p^+} + m_J \ . \tag{5.1}$$

Note that as $J^+ = [m_J + 1, m_J + l_p^+]$, the inclusion $R(J) \subseteq J^+$ follows from $R_{l_p^+} \subseteq [1, l_p^+]$. Further, for any admissible $N \in [\nu(p) + 1, \nu(p+1)]$ let

$$R_N := A_N \cup \bigcup_{J \in \mathcal{J}_N} R(J) \ . \tag{5.2}$$

Here the inclusion $R_N \subseteq [1, N]$ follows from $R(J) \subseteq J^+ \subseteq J \ \forall J \in \mathcal{J}_N$, as $J \subseteq [1, N] \ \forall J \in \mathcal{J}_N$ by definition (see Definition 5.19).

Finally, we call $j \leq N$ **regular with respect to** *N if j is contained in R_N.*

REMARK 5.23. (a) Obviously any $j \in A_N$ is regular with respect to N. As $[1, N] \setminus A_N = \Lambda_N \subseteq \Omega_\infty$ (see (5.9)), this implies that any $j \in \mathbb{N} \setminus \Omega_\infty$ is regular with respect to any $N \geq j$. In particular (see Remarks 5.12 and 5.13)

$$1, 2, l_q^+, l_q^+ + 1 \ \in \ A_N \ \subseteq \ R_N \quad \forall q \in \mathbb{N}, \ N \geq l_q^+ + 1 \ . \tag{5.3}$$

---

[1] As $l_p^+ \leq \nu(p)$ by (5.7) and l_p^+ is admissible by Remark 5.17(a).

(b) Similar to the sets A_N, the sequence $(R_N)_{N\in\mathbb{N}}$ is not increasing (compare Remark 5.15(c)). However, if $N_0 \leq N_1 \leq N_2$ are all admissible and $N_0 \in A_{N_2}$, then

$$R_{N_1} \cap [1, N_0] \;=\; R_{N_2} \cap [1, N_0] \;=\; R_{N_0}. \tag{5.4}$$

This can be seen as follows: $N_0 \in A_{N_2}$ implies that no interval $J(m)$ ($N_0 \leq m < N_2$) can reach into $[1, N_0]$, and in addition N_0 is admissible (see (5.13)). Therefore, since $R(J) \subseteq J$, all three sets in (5.4) coincide with $A_{N_0} \cup \bigcup_{J\in\mathcal{J}_{N_0}} R(J)$.

In particular, by (5.3) this implies that

$$R_N \cap [1, l_q^+] = R_{l_q^+} \text{ and } R_N \cap [1, l_q^+ + 1] = R_{l_q^+ + 1} \quad \forall N \geq l_q^+ + 1\,. \tag{5.5}$$

(c) Let $J \in \mathcal{J}$. As $R(J) = R_{l_{p_J}^+} + m_J$, statement (a) implies

$$m_J + 1, m_J + 2, m_J + l_q^+, m_J + l_q^+ + 1 \in R(J) \quad \forall q \leq p_J\,. \tag{5.6}$$

It will also be useful to have a notation for the sets of non-regular points:

DEFINITION 5.24. *For each admissible time $N \in \mathbb{N}$ let*

$$\Gamma_N \;:=\; [1, N] \setminus R_N$$

and for each $J \in \mathcal{J}$ let

$$\Gamma^+(J) \;:=\; J^+ \setminus R(J) \quad \textit{and} \quad \Gamma(J) \;:=\; J^- \cup \Gamma^+(J)\,.$$

REMARK 5.25. (a) Note that

$$\Gamma_N = \bigcup_{J\in\mathcal{J}_N} \Gamma(J) \;=\; \bigcup_{J\in\mathcal{J}_N} J^- \cup \Gamma^+(J)\,. \tag{5.7}$$

(b) Similar to (5.1), the sets $\Gamma^+(J)$ satisfy the recursive equation

$$\Gamma^+(J) \;=\; \Gamma_{l_{p_J}^+} + m_J\,. \tag{5.8}$$

(c) As $A_N \subseteq R_N$, there holds $\Gamma_N \subseteq \Lambda_N$. Thus, Remark 5.15(a) implies

$$\Gamma_N \;\subseteq\; \Lambda_N \;\subseteq\; \Omega_0 \;\subseteq\; \Omega_p \;\subseteq \Omega_\infty \tag{5.9}$$

for all admissible times $N \in \mathbb{N}$. $p \in \mathbb{N}$ is arbitrary.

(d) Suppose both N and $N + 1$ are admissible. Then $p(N) = 0$, such that $J(N) = \emptyset$, otherwise $N + 1$ would be contained in $J(N)$ could therefore not be admissible. Thus there holds $\Lambda_N = \Lambda_{N+1}$ (see (5.15)). But this means that $J_N = J_{N+1}$ and consequently $\Gamma_N = \Gamma_{N+1}$ (see (5.7)). In particular, this is true whenever $N, N + 1 \notin \Omega_\infty$, such that we obtain

$$\Gamma_{l_q^+} \;=\; \Gamma_{l_q^+ + 1} \quad \forall q \in \mathbb{N}\,. \tag{5.10}$$

Now we must gather some information about the sets R_N and Γ_N. First of all, the following lemma gives some basic control. In order to state it, let

$$\tilde{\Omega}_{-1}^{(\pm)} \;:=\; \emptyset \tag{5.11}$$

and note that $\tilde{\Omega}_0^{(\pm)} = \emptyset$ as well.

LEMMA 5.26. *(a) For any $J \in \mathcal{J}$ there holds $\Gamma(J) \subseteq \tilde{\Omega}_{p_J-2}$. Further, for any admissible $N \leq \tilde{\nu}(q)$ there holds $\Gamma_N \subseteq \tilde{\Omega}_{q-1}$.*

(b) If $j \in R(J)$ for any $J \in \mathcal{J}$, then

$$d(\omega_j, 0) \geq \frac{4\gamma}{L_2} - S_{p_J-1}(\alpha) \cdot \frac{\alpha^{-1}}{L_2} \geq \frac{3\gamma}{L_2} . \tag{5.12}$$

Further, for any admissible $N \leq \nu(q)$ there holds

$$d(\omega_j, 0) \geq \frac{4\gamma}{L_2} - S_{q-1}(\alpha) \cdot \frac{\alpha^{-1}}{L_2} \geq \frac{3\gamma}{L_2} \quad \forall j \in R_N \setminus \{N\} . \tag{5.13}$$

PROOF.

(a) We proceed by induction on q. More precisely, we prove the following induction statement:

$$\Gamma^+(J) \subseteq \tilde{\Omega}^-_{p_J-2} \quad \forall J \in \mathcal{J} : p_J \leq q \tag{5.14}$$

$$\Gamma_N \subseteq \tilde{\Omega}^-_{q-1} \quad \forall N \leq \tilde{\nu}(q) . \tag{5.15}$$

For $q = 1$ note that Γ_N is empty for all $N \leq \tilde{\nu}(1)$. In particular $\Gamma_{l_1^+}$ is empty, as $l_1^+ \leq \tilde{\nu}(1)$ by (5.7). But this means in turn that for any $J \in \mathcal{J}$ with $p_J = 1$ the set $\Gamma^+(J) = \Gamma_{l_1^+} + m_J$ is empty as well (see (5.8)).

Let $p \geq 1$ and suppose the above statements hold for all $q \leq p$. Further, let $J \in \mathcal{J}$ with $p_J = p+1$. Then $\Gamma_{l^+_{p+1}} \subseteq \tilde{\Omega}^-_{p-2}$ as $l^+_{p+1} < \tilde{\nu}(p-1)$ by (5.7). Therefore

$$\Gamma^+(J) = \Gamma_{l^+_{p+1}} + m_J \subseteq \tilde{\Omega}^-_{p-2} + m_J \subseteq \tilde{\Omega}^-_{p-1} .$$

by Lemma 5.10(b). Thus (5.14) holds for $q = p+1$.

Now suppose $N \leq \tilde{\nu}(p+1)$ and note that this implies $Q_p(m) \leq p \ \forall m < N$. Further, we have $\Gamma_N = \bigcup_{J \in \mathcal{J}_N} J^- \cup \Gamma^+(J)$ by (5.7). As $J^- \subseteq \Omega_p^-(m_J) \ \forall J \in \mathcal{J}$ and $m_J < N \ \forall J \in \mathcal{J}_N$, there holds $J^- \subseteq \tilde{\Omega}_p^-$ for any $J \in \mathcal{J}_N$, and for $\Gamma^+(J)$ the same follows from (5.14). This proves (5.15) for $q = p+1$.

(b) Suppose (5.12) holds whenever $p_J \leq p$. This implies (5.13) for all $q \leq p$: We have $d(\omega_j, 0) \geq \frac{4\gamma}{L_2}$ whenever $j \in A_N \setminus \{N\}$ for some $N \in \mathbb{N}$, and further $p_J < q \ \forall J \in \mathcal{J}_N$ whenever $N \leq \nu(q)$.

It remains to prove (5.12) by induction on p_J. If $p_J \leq 2$ the statement is obvious, because then $p(j) = 0 \ \forall j \in J \setminus \{m_J\}$ by Lemma 5.20(a).

Suppose now that (5.12) holds whenever $p_J \leq p$. As mentioned above, (5.13) then holds for all $q \leq p$. Let $p_I = p+1$ for some $I \in \mathcal{J}$ and $j \in R(I)$. Then $j - m_I \in R_{l^+_{p+1}}$ (see (5.1)), and as $l^+_{p+1} \leq \nu(p)$ we can apply (5.13) with $q = p$ to obtain that

$$d(\omega_{j-m_I}, 0) \geq \frac{4\gamma}{L_2} - S_{p-1}(\alpha) \cdot \frac{\alpha^{-1}}{L_2} .$$

Consequently

$$\begin{aligned} d(\omega_j, 0) &\geq d(\omega_{j-m_I}, 0) - d(\omega_{m_I}, 0) \geq \frac{4\gamma}{L_2} - S_{p-1}(\alpha) \cdot \frac{\alpha^{-1}}{L_2} - \frac{\alpha^{-p}}{L_2} \\ &= \frac{4\gamma}{L_2} - \left(S_{p-1}(\alpha) + \alpha^{-(p-1)}\right) \cdot \frac{\alpha^{-1}}{L_2} = \frac{4\gamma}{L_2} - S_p(\alpha) \cdot \frac{\alpha^{-1}}{L_2} . \end{aligned}$$

□

As a consequence of Lemma 5.20 and the preceding lemma, we obtain the following statements and estimates. In order to motivate these, the reader should compare the statements with the assumptions of Lemma 5.6 .

LEMMA 5.27. *(a) For any admissible $N \in \mathbb{N}$ there holds*

$$\#([1,j] \setminus R_N) \;\leq\; \frac{j}{12w} \qquad \forall j \in [1,N]\ . \tag{5.16}$$

In particular

$$\#\left([1,l_q^+] \setminus R_N\right) \;\leq\; \left[\frac{q}{12}\right] \;\leq\; \max\left\{0, \frac{2q-5}{4}\right\} \qquad \forall q \in \mathbb{N}\ , \tag{5.17}$$

where $[x]$ denotes the integer part of $x \in \mathbb{R}^+$.

(b) Let $q \geq 1$ and $\sigma := \frac{u+3}{u+v}$. Then

$$\#([j+1,l_q^+] \setminus R_{l_q^+}) \;\leq\; \sigma \cdot (l_q^+ - j) \quad \forall j \in [0, l_q^+ - 1]\ . \tag{5.18}$$

(c) Let $N \in \mathbb{N}$ be admissible, $J \in \mathcal{J}_N$ and $\lambda^+ := \lambda^+(m_J)$. Then

$$\#([j+1,\lambda^+] \cap \Gamma_N) \;\leq\; \sigma \cdot (\lambda^+ - j) \quad \forall j \in [0, \lambda^+ - 1]\ . \tag{5.19}$$

(d) Suppose $m \in \mathbb{N}$ is admissible and $p(m) \geq 1$, such that $J := J(m) \in \mathcal{J}$ by Lemma 5.21 . Then for all $q \leq p_J$ there holds

$$\#([m - l_q^-, m] \setminus R_m) \;\leq\; \frac{q}{12} \;\leq\; \max\left\{0, \frac{2q-5}{4}\right\}\ . \tag{5.20}$$

We recall that we use conditions (2.1), (2.2), (5.1) and (5.2) as standing assumptions in this subsection.

PROOF. Recall that $([1,j] \setminus R_N) = ([1,j] \cap \Gamma_N)$.

(a) This is a direct consequence of (5.9) and (5.2). For the second inequality in (5.17), note that $\#([1,l_q^+] \setminus R_N) = 0$ whenever $q < 12$.

(b) We prove the following statement by induction on q:

$$\forall j \in [0, l_q^+ - 1]\ \exists n \in [j+1, l_q^+] : \ \#[j+1,n] \cap \Gamma_{l_q^+} \;\leq\; \sigma \cdot (n-j)\ . \tag{5.21}$$

This obviously implies the statement, as it ensures the existence of a partition of $[j+1, l_q^+]$ into disjoint intervals $I_i = [j_i + 1, j_{i+1}]$ with $j = j_1 < j_2 < \ldots < j_k = l_q^+$ which all satisfy

$$\#\left(I_i \cap \Gamma_{l_q^+}\right) \;\leq\; \sigma \cdot (j_{i+1} - j_i)\ .$$

If $q = 1$, then (5.21) is obvious as $\Gamma_{l_1^+} \subseteq \Lambda_{l_1^+} = \emptyset$ (see (5.9) and note that $l_1^+ \leq \nu(1)$ by (5.7)). Now suppose (5.21) holds for all $q \leq p$. In order to show (5.21) for $p+1$, we have to distinguish three possible cases. Recall that by (5.7) and (5.9)

$$\Gamma_{l_{p+1}^+} \;=\; \bigcup_{J \in \mathcal{J}_{l_{p+1}^+}} J^- \cup \Gamma^+(J) \;\subseteq\; \Lambda_{l_{p+1}^+}\ .$$

If $j+1 \notin \Gamma_{l_{p+1}^+}$ we can choose $n = j+1$.

If $j+1 \in \Gamma^+(J)$ for some $J \in \mathcal{J}_{l^+_{p+1}}$ then $p_J \le p$ as $l^+_{p+1} < \nu(p)$ by (5.7). By (5.8) there holds $j - m_J \in \Gamma_{l^+_{p_J}} \subseteq [0, l^+_{p_J} - 1]$. Thus we can apply the induction statement with $q = p_J$ to $j - m_J$ and obtain some $\tilde{n} \in [j - m_J + 1, l^+_{p_J}]$ with

$$\#([j - m_J + 1, \tilde{n}] \cap \Gamma_{l^+_{p_J}}) \;\le\; \sigma \cdot (\tilde{n} - j + m_J)\;.$$

As $\Gamma^+(J) = \Gamma_{l^+_{p_J}} + m_J$ (again by (5.8)), $n := \tilde{n} + m_J$ has the required property.

Finally, if $j+1 \in J^-$ for some $J \in \mathcal{J}_{l^+_{p+1}}$ then $[\lambda^-(m_J), j+1] \subseteq J^- \subseteq \Gamma_{l^+_{p+1}}$. Therefore

$$\text{(5.22)}\qquad \begin{aligned} \frac{\#\left([j+1, \lambda^+(m_J)] \cap \Gamma_{l^+_{p+1}}\right)}{\lambda^+(m_J) - j} &\le \frac{\#\left(J \cap \Gamma_{l^+_{p+1}}\right)}{\#J} \\ &\le \frac{\#(J^- \cup \Gamma^+(J))}{(u+v)\cdot p_J} \le \frac{(u+2)\cdot p_J + \#\Gamma_{l^+_{p_J}}}{(u+v)\cdot p_J} \le \frac{u+3}{u+v}\,. \end{aligned}$$

where we used part (a) of this lemma with $j = N = l^+_{p_J}$ to conclude that $\#\Gamma_{l^+_{p_J}} \le p_J$.

(c) Similar to (a), we prove that

$$\forall j \in [0, \lambda^+ - 1]\ \exists n \in [j+1, \lambda^+] :\ \#([j+1, n] \cap \Gamma_N) \;\le\; \sigma(n-j)\;.$$

Again, we have to distinguish three cases:

If $j+1 \notin \Gamma_N$ we can choose $n = j+1$.

If $j+1 \in \Gamma^+(I)$ for some $I \in \mathcal{J}_N$, then we can choose $n = \lambda^+(m_I) = m_I + l^+_{p_I}$. Using that $\Gamma^+(I) = \Gamma_{l^+_{p_I}} + m_I$ by (5.8), part (b) implies that n has the required property.

If $j+1 \in I^-$ for some $I \in \mathcal{J}_N$ we can choose $n = \lambda^+(m_I)$ and proceed exactly as in (5.22), with J being replaced by I.

(d) By Lemma 5.20(b) there holds $m - l^-_q \in A_m$. Therefore (5.7), $\Gamma(J) \subseteq J$ and (5.8) imply that

$$[m - l^-_q, m] \cap \Gamma_m \;\subseteq\; \bigcup_{j \in [m - l^-_q + 1, m-1]} \Omega_{q-2}(j) \;=:\; U\;.$$

Further, Lemma 5.20(b) yields that $U \subseteq \tilde{\Omega}_{q-2}$, such that $U - m \subseteq \Omega_\infty$ by Lemma 5.10(b). Consequently

$$\begin{aligned} &\#\left([m - l^-_q, m] \cap \Gamma_m\right) \;\le\; \#U \;=\; \#(U - m) \\ &\le\; \#\left([-l^-_q, -1] \cap \Omega_\infty\right) \overset{(5.2)}{\le} \frac{l^-_q}{12w} \le \frac{q}{12} \le \max\left\{0, \frac{2q-5}{12}\right\}\,. \end{aligned}$$

□

CHAPTER 6

Construction of the sink-source orbits: One-sided forcing

We now turn to the construction of the sink-source-orbits in the case of one-sided forcing. Before we start with the core part of the proof, we have to add some more assumptions on the parameters. Further, we restate two estimates from the preceding section, together with a few other facts that will be used frequently in the construction.

First of all, we choose u and v such that

$$u \geq 8 , \tag{6.1}$$

$$v \geq 8 , \tag{6.2}$$

$$\sigma \leq \tfrac{1}{6} . \tag{6.3}$$

In addition, we assume that

$$\tfrac{1}{2}\sqrt{\alpha} \geq 6 + K \cdot S_\infty(\alpha^{\frac{1}{4}}) . \tag{6.4}$$

Further, we remark that (2.2) implies

$$\alpha \geq 4S_\infty(\alpha) . \tag{6.5}$$

Now suppose that (2.1), (2.2), (5.1) and (5.2) hold. Together with the above assumptions and the respective results from the last section (see (5.7), Lemma 5.27(b) and (5.5)), this yields that for any $q \geq 1$ the following estimates hold:

$$4(q+1) \leq 8q \leq l_q^\pm < \tilde{\nu}(\max\{1, q-2\}) \leq \nu(\max\{1, q-2\}) \tag{6.6}$$

$$\#([j+1, l_q^+] \setminus R_N) \leq \frac{l_q^+ - j}{6} \quad \forall N \geq l_q^+,\ j \in [0, l_q^+ - 1] . \tag{6.7}$$

Recall that $(\xi_n(\beta, l))_{n \geq -l}$ corresponds the forward orbit of the point $(\omega_{-l}, 3)$ under the transformation T_β, where we suppress the θ-coordinate (see Definition 4.1). As we are in the case of one-sided forcing, we can use the fact that for all $l, n \in \mathbb{Z}, n \geq -l$ the mapping $\beta \mapsto \xi_n(\beta, l)$ is monotonically decreasing in β. For $l \geq 0$ and $n \geq 1$ the monotonicity is even strict (as $g(0) = 1 > 0$ and F is strictly increasing by (2.5)). This has some very convenient implications. First of all, we can uniquely define parameters $\beta_{q,n}^+$ and $\beta_{q,n}^-$ $(q, n \in \mathbb{N})$ by the equations

$$\xi_n(\beta_{q,n}^+, l_q^-) = \frac{1}{\alpha} \tag{6.8}$$

and

$$\xi_n(\beta_{q,n}^-, l_q^-) = -\frac{1}{\alpha} . \tag{6.9}$$

In addition, we let

$$l_0^- := 0 \quad \text{and} \quad l_0^+ := 0 \tag{6.10}$$

(note that so far the $l_q^\pm$ had only been defined for $q \geq 1$) and extend the definitions of $\beta_{q,n}^\pm$ to $q = 0$. If we now want to show that $\xi_n(\beta, l_q^-) \in \overline{B_{\frac{1}{\alpha}}(0)}$ implies $\xi_j(\beta, l_q^-) \in \overline{B_{\frac{1}{\alpha}}(0)}$ for some $j < n$, we can do so by proving that

$$\xi_n(\beta_{q,j}^+, l_q^-) \geq \frac{1}{\alpha} \tag{6.11}$$

and

$$\xi_n(\beta_{q,j}^-, l_q^-) \leq -\frac{1}{\alpha} \tag{6.12}$$

(compare (4.2)–(4.5)). Furthermore, (6.12) is a trivial consequence of the fact that $\mathbb{T}^1 \times [-3, -\frac{1}{\alpha}]$ is mapped into $\mathbb{T}^1 \times [-3, -(1-\gamma)] \subseteq \mathbb{T}^1 \times [-3, -\frac{1}{\alpha})$ (see (2.8)). Thus, it always suffices to consider (6.11).

Now we can formulate the induction statement we want to prove:

INDUCTION SCHEME 6.1. *Suppose the assumptions of Theorem 2.7 are satisfied and (5.1), (5.1), (5.2) and (6.1)–(6.4) hold. Then for any $q \in \mathbb{N}_0$ there holds*

I. *If $\xi_{l_q^+ + 1}(\beta, l_q^-) \in \overline{B_{\frac{1}{\alpha}}(0)}$ then*

$$\xi_j(\beta, l_q^-) \geq \gamma \quad \forall j \in [-l_q^-, 0] \setminus \Omega_\infty \tag{6.13}$$

and $\beta \in \left[1 + \frac{1}{\sqrt{\alpha}}, 1 + \frac{3}{\sqrt{\alpha}}\right]$.

II. *Suppose $n \in [l_q^+ + 1, \nu(q+1)]$ is admissible. Then $\xi_n(\beta, l_q^-) \in \overline{B_{\frac{1}{\alpha}}(0)}$ implies that (6.13) holds,*

$$\xi_j(\beta, l_q^-) \in \overline{B_{\frac{1}{\alpha}}(0)} \qquad \forall j \in R_n \tag{6.14}$$

and $\beta \in \left[1 + \frac{1}{\sqrt{\alpha}}, 1 + \frac{3}{\sqrt{\alpha}}\right]$.

III. *Let $1 \leq q_1 \leq q$ and suppose $n_1 \in [l_{q_1}^+ + 1, \nu(q_1 + 1)]$ and $n_2 \in [l_q^+ + 1, \nu(q+1)]$ are both admissible.*

(a) If $q_1 = q$ and $n_1 \in R_{n_2}$, then

$$|\beta_{q_1,n_1}^+ - \beta_{q,n_2}^+| \leq 2\alpha^{-\frac{n_1}{4}} . \tag{6.15}$$

(b) If $q_1 < q$ there holds

$$|\beta_{q_1,n_1}^+ - \beta_{q,n_2}^+| \leq 3 \cdot \sum_{i=q_1+1}^{q} \alpha^{-i} \leq \alpha^{-q_1} . \tag{6.16}$$

The proof is given in the next subsection. The statement of Theorem 2.7 now follows easily, with the help of Lemma 2.6:

Proof of Theorem 2.7 . In order to apply Lemma 2.6 we can use the same sequences $l_p^\pm$ as in Induction Scheme 6.1. Further, let $\beta_p := \beta_{p,l_p^+ + 1}^+$, $\theta_p := \omega$ and $x_p := \xi_1(\beta_p, l_p^-)$. From Part II of the induction statement with $q = p$ and $n = l_p^+ + 1$ we obtain that

$$\xi_j(\beta_p, l_p^-) \in \overline{B_{\frac{1}{\alpha}}(0)} \qquad \forall j \in R_{l_p^+ + 1} ,$$

and Lemma 5.27(a) implies that

$$\#\left([1,j]\cap R_{l_p^+ +1}\right) \;\geq\; \frac{11}{12}\cdot j \qquad \forall j\in[1,l_p^+]\ .$$

Therefore it follows from (2.5) and (2.6) that

$$\lambda^+(\beta_p,\theta_p,x_p,j) \;=$$

$$=\; \frac{1}{j}\sum_{i=1}^{j}\log F'(\xi_i(\beta_p,l_p^-)) \;\geq\; \frac{11}{12}\cdot\frac{\log\alpha}{2}-\frac{2\log\alpha}{12} \;=\; \frac{7}{24}\cdot\log\alpha \qquad \forall j\in[1,l_p^+]\ .$$

Likewise, we can conclude from Part I of the induction statement with $q=p$ in combination with (5.2), (2.5) and (2.7) that

$$\lambda^-(\beta_p,\theta_p,x_p,j) \;\geq\; \frac{7}{24}\cdot\log\alpha \qquad \forall j\in[1,l_p^-]\ .$$

Consequently, the assertions of Lemma 2.6 are satisfied, such that there is at least one parameter value at which a sink-source-orbit and consequently an SNA and an SNR occur (see Theorem 2.4). Due to Theorem 2.1, the only parameter where this is possible is the critical parameter β_0. Finally the statement about the essential closure again follows from Theorem 2.1 .

□

Proof of Addendum 2.9. Define β_p, θ_p and x_p as above. From Part III of the induction statement it follows that $(\beta_p)_{p\in\mathbb{N}}$ is a Cauchy-sequence and must therefore converge to β_0 (instead of only having a convergent subsequence, as in the proof of Lemma 2.6). To be more precise, if $p<q$ we have $|\beta_p-\beta_q|\leq\alpha^{-p}$, such that

$$|\beta_p-\beta_0| \;\leq\; \alpha^{-p} \qquad \forall p\in\mathbb{N}\ . \tag{6.17}$$

Further, let

$$\theta_0 \;:=\; \omega$$

and

$$x_0 \;:=\; \lim_{p\to\infty} x_p\ . \tag{6.18}$$

If the limit in (6.18) does not exist,[1] we just go over to a suitable subsequence. From Part II of the induction statement with $q=p$ and $n=l_p^+ +1$, it follows that

$$T_{\beta_p,\omega,j-1}(x_p) \;=\; \xi_j(\beta_p,l_p^-) \in \overline{B_{\frac{1}{\alpha}}(0)} \qquad \forall j\in R_{l_p^+ +1}\ .$$

Using that $R_{l_p^+ +1}\subseteq R_{l_q^+ +1}$ $\forall q\geq p$ by (5.5) and the continuity of the map $(\beta,x)\mapsto T_{\beta,\omega,j-1}(x)$, we see that

$$T_{\beta_0,\omega,j-1}(x_0) \;\in\; \overline{B_{\frac{1}{\alpha}}(0)} \qquad \forall j\in R_{l_p^+ +1}, p\in\mathbb{N}\ . \tag{6.19}$$

Now φ^+ and ψ can be defined pointwise as the upper bounding graph (see (1.4) of the system T_{β_0} and by equation (3.1), respectively. Then the fact that

$$\psi(\omega) \;\leq\; x_0 \tag{6.20}$$

[1] In fact it is possible to show that $(x_p)_{p\in\mathbb{N}}$ is a Cauchy-sequence as well, by using Lemma 5.2 and Part I of the induction statement. However, we refrain from doing so as this is not relevant for the further argument.

is obvious, otherwise the forward orbit of (θ_0, x_0) would converge to the lower bounding graph φ^- and its forward Lyapunov exponent would therefore be negative. On the other hand suppose

$$\psi(\omega) \geq x_0 - \alpha^{-p}$$

for some $p \geq 2$. Then we can compare the orbits

$$x_1^1, \ldots, x_n^1 := x_0, \ldots, T_{\beta_0, \omega_1, l_p^+ - 1}(x_0) \tag{6.21}$$

and

$$x_1^2, \ldots, x_n^2 := \psi(\omega_1), \ldots, \psi(\omega_{l_p^+}) \tag{6.22}$$

via Lemma 5.6(b)[2] and obtain that $\psi(\omega_{l_p^+ + 1}) \leq -\frac{2}{\alpha}$. But as we have seen in the proof of Theorem 2.1 that all points below the 0-line eventually converge to the lower bounding graph, this contradicts the definition of ψ. Consequently

$$x_0 \leq \psi(\omega) + \alpha^{-p} \quad \forall p \in \mathbb{N} .$$

Together with (6.20) this implies that $x_0 = \psi(\omega)$.

As $\psi \leq \varphi^+$, we immediately obtain $x_0 \leq \varphi^+(\omega)$, such that it remains to show

$$x_0 \geq \varphi^+(\omega) . \tag{6.23}$$

To that end, we denote the upper boundary lines of the system (2.1) by φ_n if $\beta = \beta_0$ and by $\varphi_{p,n}$ if $\beta = \beta_p$. Now either infinitely many β_p are below β_0, or infinitely many β_p are above β_0. Therefore, by going over to a suitable subsequence if necessary, we can assume w.l.o.g. that either $\beta_p \leq \beta_0 \ \forall p \in \mathbb{N}$ or $\beta_p \geq \beta_0 \ \forall p \in \mathbb{N}$. The first case is treated rather easily: If $\beta_p \leq \beta_0$, then

$$x_p = \xi_1(\beta_p, l_p^-) = \varphi_{p, l_p^- + 1}(\omega) \geq \varphi_{l_p^- + 1}(\omega) \geq \varphi^+(\omega) .$$

Passing to the limit $p \to \infty$, this proves (6.23).

On the other hand, suppose $\beta^p \geq \beta_0$. In this case, we will show that

$$|x_p - \varphi_{l_p^- + 1}(\omega)| = |\xi_1(\beta_p, l_p^-) - \xi_1(\beta_0, l_p^-)| \leq \alpha^{-p} \cdot \left(6 + K \cdot S_\infty(\alpha^{\frac{1}{4}})\right) . \tag{6.24}$$

As $\varphi_n(\omega) \stackrel{n\to\infty}{\longrightarrow} \varphi^+(\omega)$ and $x_p \stackrel{p\to\infty}{\longrightarrow} x_0$, this again proves (6.23). Note that as $\beta_p \geq \beta_0$ we have $\xi_j(\beta_0, l_p^-) \geq \xi_j(\beta_p, l_p^-) \ \forall j \geq -l_p^-$, such that $\xi_j(\beta_p, l_p^-) \geq \gamma$ implies $\xi_j(\beta_0, l_p^-) \geq \gamma$. This allows to compare the orbits

$$x_1^1, \ldots, x_n^1 := \xi_{-l_p^-}(\beta_p, l_p^-), \ldots, \xi_0(\beta_p, l_p^-) \tag{6.25}$$

and

$$x_1^1, \ldots, x_n^1 := \xi_{-l_p^-}(\beta_0, l_p^-), \ldots, \xi_0(\beta_0, l_p^-) \tag{6.26}$$

via Lemma 5.2,[3] which yields (6.24).

---

[2]We can choose $\epsilon = \frac{\alpha^{-p}}{2}$, such that $q = p$. Note that the error term is zero, as we consider orbits which are located on the same fibres and generated with the same parameter. As $l_p^+ + 1 \in R_{l_p^+ + 1}$, $x_{n+1}^1 \in B_{\frac{1}{\alpha}}(0)$ follows from (6.19). $\tau(n) \leq \frac{2p-3}{4}$ and $\tau(j) \leq \frac{j}{8}$ follow from Lemma 5.27(a), whereas $\tau(n) - \tau(j) \leq \frac{n-j}{6}$ follows from (6.7). Finally $n = l_p^+ \geq 5p$ by (6.6).

[3]With $\epsilon = \alpha^{-p}$. We have $|\beta_p - \beta_0| \leq \alpha^{-p}$ by (6.17), such that $\mathrm{err}(\ldots) \leq \epsilon$. $\eta(j,n) \leq \frac{n+1-j}{10}$ follows from Part I of the induction statement with $q = p$ together with (5.2) and $0 \notin \Omega_\infty$. Finally $n = l_p^- + 1 \geq 4p$ by (6.6), such that $\alpha^{-\frac{n}{4}} \leq \epsilon$.

□

6.1. Proof of the induction scheme

Standing assumption: In this whole subsection, we always assume that the assumptions of the Induction scheme 6.1 are satisfied.

Before we start the proof of the Induction statement, we provide the following lemma, which will be used in order to obtain estimates on the parameters $\beta^{+}_{q,n}$:

LEMMA 6.2. *Suppose Let n be admissible and $\xi_n(\beta_1, l), \xi_n(\beta_2, l) \in \overline{B_{\frac{1}{\alpha}}(0)}$. Further, suppose that $\xi_n(\beta, l) \in \overline{B_{\frac{1}{\alpha}}(0)}$ implies $\xi_j(\beta, l) \in \overline{B_{\frac{1}{\alpha}}(0)}$ $\forall j \in R_n$. Then*

$$|\beta_1 - \beta_2| \;\leq\; 2\alpha^{-\frac{n}{4}} \; .$$

PROOF. Note that

$$\frac{\partial}{\partial \beta}\xi_{j+1}(\beta, l) \;=\; \frac{\partial}{\partial \beta}\left(F(\xi_j(\beta, l)) - \beta \cdot g(\omega_j)\right) \tag{6.1}$$

$$= \;\; F'(\xi_j(\beta, l)) \cdot \frac{\partial}{\partial \beta}\xi_j(\beta, l) - g(\omega_j) \;\stackrel{(g \geq 0)}{\leq}\; F'(\xi_j(\beta, l)) \cdot \frac{\partial}{\partial \beta}\xi_j(\beta, l) \; .$$

W.l.o.g. we can assume $\beta_1 < \beta_2$. As we have $\frac{\partial}{\partial \beta}\xi_0(\beta, k) \leq -1$, the inductive application of (6.1) together with (2.5) and (2.6) yields

$$\frac{\partial}{\partial \beta}\xi_n(\beta, l) \;\leq\; -(\alpha^{\frac{1}{2}})^{\#([1,n-1]\cap R_n])} \cdot (\alpha^{-2})^{\#([1,n-1]\setminus R_n)} \;=\; -\alpha^{\frac{1}{2}(n-1-5\cdot\#\Gamma_n)}$$

as long as $\xi_n(\beta, k) \in \overline{B_{\frac{1}{\alpha}}(0)}$. (Recall that $[1, n] \setminus R_n = \Gamma_n$ by definition and $n \in R_n$ by assumption.) In particular this is true for all $\beta \in [\beta_1, \beta_2]$. Lemma 5.27(a) yields $\#\Gamma_n = \#([1, n-1] \setminus R_n) \leq \frac{n-1}{10}$, such that we obtain

$$\frac{\partial}{\partial \beta}\xi_n(\beta, l) \;\leq\; -\alpha^{\frac{n-1}{4}}$$

The required estimate now follows from $|\xi_n(\beta_1, l) - \xi_n(\beta_2, l)| \leq \frac{2}{\alpha}$.

□

We prove the Induction scheme 6.1 by induction on q, proceeding in six steps. The first one starts the induction:

Step 1: *Proof of the statement for $q = 0$.*

As $d(\omega_j, 0) \geq \frac{4\gamma}{L_2}$ $\forall j \in [1, \nu(1) - 1]$, Part I and II of the induction statement are already contained in Lemma 4.2, and Part III is still void.

■

Now let $p \geq 1$ and assume that the statement of Induction scheme 6.1 holds for all $q \leq p - 1$. We have to show that the statement then holds for p as well. The next two steps will prove Part I of the induction statement for p. Note that for $p = 1$ Part I of the induction statement is still contained in Lemma 4.2 as $l_1^{\pm} < \nu(1)$ by (6.6). Therefore, we can assume

$$p \geq 2 \tag{6.2}$$

during Step 2 and Step 3.

Step 2: *If* $|\beta - \beta^+_{p-1,\nu(p)}| \leq \alpha^{-p}$, *then* $\xi_j(\beta, l^-_p) \geq \gamma \ \forall j \in [-l^-_p, 0] \setminus \Omega_\infty$.

This is a direct consequence of the following lemma with $q = p$, $l^* = l^-_{p-1}$, $l = l^-_p$, $\beta^* = \beta^+_{p-1,\nu(p)}$, $m = \nu(p)$ and $k = -\nu(p)$, .[4] Note that $\tilde{\Omega}_{p-2} - \nu(p) \subseteq \tilde{\Omega}_{p-1} \subseteq \Omega_\infty$ by Lemma 5.10(b). The statement of the lemma is slightly more general because we also want to use it in similar situations later. Recall that $\tilde{\Omega}_{-1} = \tilde{\Omega}_0 = \emptyset$, see (5.11).

LEMMA 6.3. *Let* $q \geq 1$, $l^*, l \geq 0$, $\beta^* \in \left[1 + \frac{1}{\sqrt{\alpha}}, 1 + \frac{3}{\sqrt{\alpha}}\right]$ *and* $|\beta - \beta^*| \leq 2\alpha^{-q}$. *Suppose that* m *is admissible,* $p(m) \geq q$ *and either* $k = 0$ *or* $p(k) \geq q$. *Further, suppose*

$$\xi_j(\beta^*, l^*) \ \in \ \overline{B_{\frac{1}{\alpha}}(0)} \quad \forall j \in R_m$$

and $\xi_{m+k-l^-_q}(\beta, l) \geq \gamma$. *Then*

$$\{j \in [m - l^-_q, m] \mid \xi_{j+k}(\beta, l) < \gamma\} \ \subseteq \ \tilde{\Omega}_{q-2} \, .$$

PROOF. We have that $J(m) \in \mathcal{J}$ by Lemma 5.21, such that we can apply Lemma 5.20(b) to $J := J(m)$. Note that $m = m_J$ and $p_J = p(m) \geq q$ in this case. Let $t := m - l^-_q$. We will show that

$$\{j \in [t, m] \mid \xi_{j+k}(\beta, l) < \gamma\} \ \subseteq \ \bigcup_{t \leq j < m} [\lambda^-(j), \lambda^+(j) + 1] \, .$$

As $[\lambda^-(j), \lambda^+(j) + 1] \subseteq \Omega_0(j) \subseteq \Omega_{q-2}(j)$ and $Q_{q-2}(j) \leq Q_\infty(j) \leq \max\{0, q - 2\} \ \forall j \in [t, m-1]$ (see Lemma 5.20(b)), this proves the statement.

Let $J_1, \ldots, J_r$ be the ordered sequence of intervals $J \in \mathcal{J}_m$ with $J \subseteq [t, m]$, such that

$$[t, m] \setminus A_m \ = \ [t, m] \cap \Lambda_m \ = \ \bigcup_{i=1}^{r} J_i$$

(recall that $[1, m] \setminus A_m = \Lambda_m$ by definition). Further, define

$$j^-_i \ := \ \lambda^-(m_{J_i}) \quad \text{and} \quad j^+_i \ := \ \lambda^+(m_{J_i}) \, ,$$

such that $J_i = [j^-_i, j^+_i]$. We have to show that $\xi_{j+k}(\beta, l) \geq \gamma$ whenever j is contained in $[j_i + 2, j^-_{i+1} - 1]$ for some $i = 1, \ldots, r$ or in $[t, j^-_1 - 1] \cup [j^+_r + 2, m]$, and we will do so by induction on i. The case where $j^+_i + 1 = j^-_{i+1} - 1$, such that $[j^+_i + 2, j^-_{i+1} - 1]$ is empty, is somewhat special and will be addressed later, so for now we always assume $j^+_i + 1 < j^-_{i+1} - 1$.

Let us first see that $\xi_{j^+_i+2+k}(\beta, l) \geq \gamma$ implies $\xi_{j+k}(\beta, l) \geq \gamma \ \forall j \in [j^+_i + 2, j^-_{i+1} - 1]$

---

[4]Note that $\nu(p)$ is admissible by Lemma 5.20(d). Therefore $\beta^* = \beta^+_{p-1,\nu(p)} \in \left[1 + \frac{1}{\sqrt{\alpha}}, 1 + \frac{3}{\sqrt{\alpha}}\right]$ and $\xi_j(\beta^*, l^-_{p-1}) \in \overline{B_{\frac{1}{\alpha}}(0)} \ \forall j \in R_{\nu(p)}$ follow from Part II of the induction statement with $q = p - 1$ and $n = \nu(p)$.

: If $j \in [j_i^+ + 2, j_{i+1}^- - 1]$, then $j \in A_m$. Hence $d(\omega_j, 0) \geq \frac{4\gamma}{L_2}$ and therefore

$$d(\omega_{j+k}, 0) \geq \frac{4\gamma - \alpha^{-q}}{L_2} \geq \frac{3\gamma}{L_2}$$

by (2.2) if $q \geq 2$. In case $q = 1$ we obtain the same result, as $\tilde{\nu}(1) > l_1^-$ by (6.6) then implies that $d(\omega_j, 0) \geq \frac{8\gamma}{L_2}$ $\forall j \in [t, m]$. Further $\beta \in [1, 1 + \frac{4}{\sqrt{\alpha}}]$ as $|\beta - \beta^*| \leq 2\alpha^{-q} \leq \frac{2}{\alpha} \overset{(2.2)}{\leq} \frac{1}{\sqrt{\alpha}}$ and $\beta^* \in [1 + \frac{1}{\sqrt{\alpha}}, 1 + \frac{3}{\sqrt{\alpha}}]$. Inductive application of Lemma 4.3 therefore yields

$$\xi_{j+k}(\beta, l) \geq \gamma \quad \forall j \in [j_i^+ + 2, j_{i+1}^- - 1] \, .$$

The same argument also starts and ends the induction: As $\xi_{t+k}(\beta, l) \geq \gamma$ by assumption we get $\xi_{j+k}(\beta, l) \geq \gamma$ $\forall j \in [t, j_1^- - 1]$, and for $j \in [j_r^+ + 2, m]$ this follows from $\xi_{j_r^+ + 2 + k}(\beta, l) \geq \gamma$.

If $q = 1$, then Lemma 5.20(b) yields that $p(j) = 0$ $\forall j \in [t, m]$ and consequently $[t, m] \setminus A_m = \emptyset$, such that we are already finished in this case. Therefore, we can assume from now on that $q \geq 2$. It remains to prove that

(6.3) $$\xi_{j_i^- - 1 + k}(\beta, l) \geq \gamma \quad \text{implies} \quad \xi_{j_i^+ + 2 + k}(\beta, l) \geq \gamma \, .$$

In order to do this, we have to apply Lemma 5.6(a): Let $\epsilon := \alpha^{-(q-1)}$ and choose

(6.4) $$x_1^1, \ldots, x_n^1 \; := \; \xi_{j_i^- - 1}(\beta^*, l^*), \ldots, \xi_{j_i^+}(\beta^*, l^*)$$

and

(6.5) $$x_1^2, \ldots, x_n^2 \; := \; \xi_{j_i^- - 1 + k}(\beta, l), \ldots, \xi_{j_i^+ + k}(\beta, l) \, .$$

As $d(\omega_k, 0) \leq \frac{\alpha^{-(q-1)}}{L_2}$ and $|\beta - \beta^*| \leq 2\alpha^{-q}$ we have $\text{err}(\beta_1, \beta_2, \theta_1, \theta_2) \leq K \cdot \epsilon$ by Remark 4.5 . Further $x_1^1 = \xi_{j_i^- - 1}(\beta^*, l^*) \in \overline{B_{\frac{1}{\alpha}}(0)}$ and $x_{n+1}^1 = \xi_{j_i^+ + 1}(\beta^*, l^*) \in \overline{B_{\frac{1}{\alpha}}(0)}$ by assumption (as $j_i^- - 1, j_i^+ + 1 \in A_m \subseteq R_m$), whereas $x_1^2 = \xi_{j_i^- - 1 + k}(\beta, l) \geq \gamma \geq \frac{2}{\alpha}$. Applying Lemma 5.27(d) we obtain that

$$\tau(n) = \#([j_i^- - 1, j_i^+] \setminus R_m) \leq \#([t, m] \setminus R_m) \overset{(5.20)}{\leq} \min\left\{0, \frac{2q - 5}{4}\right\}$$

Finally, we have

$$|\tau(n) - \tau(j)| \leq \#([j_i^+ - (n - j) + 1, j_i^+]) \setminus R_m \; \leq \; -\sigma \cdot (n - j) \leq \frac{n - j}{6}$$

by Lemma 5.27(c) (with $N = m$, $J = J_i$ and $\lambda^+ = \lambda^+(m_{J_i}) = j_i^+$). Thus all the assumptions of Lemma 5.6 are satisfied and we can conclude that $x_{n+1}^2 = \xi_{j_i^+ + 1 + k}(\beta, l) \geq \frac{2}{\alpha}$. As we have $d(\omega_{j_i^+ + 1 + k}, 0) \geq \frac{3\gamma}{L_2}$ again, Lemma 4.3 now implies $\xi_{j_i^+ + 2 + k}(\beta, l) \geq \gamma$.

As mentioned, we still have to address the case where $j_i^+ + 1 = j_{i+1}^- - 1$, such that $[j_i^+ + 2, j_{i+1}^- - 1]$ is empty. In this case we still obtain that $\xi_{j_i^+ + 1 + k}(\beta, l) = \xi_{j_{i+1}^- - 1 + k}(\beta, l) \geq \frac{2}{\alpha}$. But this is sufficient in order to apply Lemma 5.6(a) once more, in exactly the same way as above, to conclude that $\xi_{j_{i+1}^+ + 1 + k}(\beta, l) \geq \frac{2}{\alpha}$. Thus in the next step we obtain $\xi_{j_{i+1}^+ + 2 + k}(\beta, l) \geq \gamma$ as before, unless again $j_{i+1}^+ + 1 = j_{i+2}^- - 1$. In any case, the induction can be continued.

$\square$

$\blacksquare$

Step 3: $\xi_{l_p^+ +1}(\beta, l_p^-) \in \overline{B_{\frac{1}{\alpha}}(0)}$ *implies* $|\beta - \beta^+_{p-1,\nu(p)}| \leq \alpha^{-p}$.

Recall that we can assume $p \geq 2$, see (6.2). Let $\beta^* := \beta^+_{p-1,\nu(p)}$, $\beta^+ := \beta^* - \alpha^{-p}$ and $\beta^- := \beta^* + \alpha^{-p}$. We prove

CLAIM 6.4. $$\xi_{l_p^+ +1}(\beta^+, l_p^-) > \tfrac{1}{\alpha} \quad .$$

As $\xi_{l_p^+ +1}(\beta^-, l_p^-) < -\frac{1}{\alpha}$ follows in exactly the same way, this implies the statement.

Proof of the claim:
Using Step 2, we see that

$$\xi_j(\beta^+, l_p^-) \ \geq\ \gamma \qquad \forall j \in [-l_p^-, 0] \setminus \Omega_\infty \ . \tag{6.6}$$

On the other hand, from Part II of the the induction statement with $q = p-1$ and $n = \nu(p)$ it follows that[5]

$$\xi_j(\beta^*, l_{p-1}^-) \ \geq\ \gamma \qquad \forall j \in [-l_{p-1}^-, 0] \setminus \Omega_\infty \tag{6.7}$$

and

$$\xi_j(\beta^*, l_{p-1}^-) \ \in\ \overline{B_{\frac{1}{\alpha}}(0)} \qquad \forall j \in R_{\nu(p)} \ . \tag{6.8}$$

Thus we can use Lemma 5.2 with $\epsilon = \alpha^{-p}$ to compare the sequences

$$x_1^1, \ldots, x_n^1 \ :=\ \xi_{-l_{p-1}^-}(\beta^*, l_{p-1}^-), \ldots, \xi_{-1}(\beta^*, l_{p-1}^-) \tag{6.9}$$

and

$$x_1^2, \ldots, x_n^2 := \xi_{-l_{p-1}^-}(\beta^+, l_p^-), \ldots, \xi_{-1}(\beta^+, l_p^-) \tag{6.10}$$

and obtain that[6]

$$|\xi_0(\beta^+, l_p^-) - \xi_0(\beta^*, l_{p-1}^-)| \ \leq\ \alpha^{-p} \cdot (6 + K \cdot S_\infty(\alpha^{-\frac{1}{4}})) \ .$$

Note that (6.6) and (6.7) in particular imply that both $\xi_0(\beta^*, l_{p-1}^-) \geq \gamma$ and $\xi_0(\beta^+, l_p^-) \geq \gamma$. Therefore we can use (2.7) to obtain

$$\begin{aligned} \xi_1(\beta^+, l_p^-) \quad &\geq \quad \xi_1(\beta^*, l_{p-1}^-) + (\beta^* - \beta^+) - \alpha^{-p} \cdot \frac{6 + K \cdot S_\infty(\alpha^{\frac{1}{4}})}{2\sqrt{\alpha}} \\ &\overset{(6.4)}{\geq} \quad \xi_1(\beta^*, l_{p-1}^-) + \frac{\alpha^{-p}}{2} \ . \end{aligned}$$

Now we compare

$$x_1^1, \ldots, x_n^1 \ :=\ \xi_1(\beta^*, l_{p-1}^-), \ldots, \xi_{l_p^+}(\beta^*, l_{p-1}^-) \tag{6.11}$$

---

[5]Note that $\nu(p)$ is admissible by Lemma 5.20(d) and $\xi_{\nu(p)}(\beta^*, l_{p-1}^-) \in \overline{B_{\frac{1}{\alpha}}(0)}$ by definition of $\beta^* = \beta^+_{p-1,\nu(p)}$.

[6]As the two orbits lie on the same fibres and $\beta^* - \beta^+ = \alpha^{-p}$, we have $\mathrm{err}(\ldots) \leq K \cdot \epsilon$, see Remark 4.5 . Further, by (6.6) and (6.7) we have $\eta(j,n) \leq \#([-(n-j), -1] \cap \Omega_\infty) \leq \frac{n+1-j}{10}$ by (5.2) and $n = l_{p-1}^- \geq 4p$ by (6.6).

and

$$x_1^2, \dots, x_n^2 := \xi_1(\beta^+, l_p^-), \dots, \xi_{l_p^+}(\beta^+, l_p^-) \tag{6.12}$$

via Lemma 5.6(b)[7] and obtain that $\xi_{l_p^+}(\beta^+, l_p^-) = x_{n+1}^2 \geq \frac{2}{\alpha}$.

□

■

Step 2 and 3 together prove Part I of the induction statement for $q = p$, apart from $\beta \in \left[1 + \frac{1}{\sqrt{\alpha}}, 1 + \frac{3}{\sqrt{\alpha}}\right]$ whenever $\xi_{l_p^+ +1}(\beta, l_p^-) \in \overline{B_{\frac{1}{\alpha}}(0)}$. This will be postponed until after the next step. However, Step 3 implies the slightly weaker estimate

$$\beta \in \left[1 + \frac{1}{\sqrt{\alpha}} - \alpha^{-p}, 1 + \frac{3}{\sqrt{\alpha}} + \alpha^{-p}\right] .$$

(Note that as $\nu(p)$ is admissible the induction statement can be applied to $q = p-1$ and $n = \nu(p)$, such that $\beta^+_{p-1,\nu(p)} \in \left[1 + \frac{1}{\sqrt{\alpha}}, 1 + \frac{3}{\sqrt{\alpha}}\right]$.) This will be sufficient in the meanwhile.

The next three steps will prove Part II and III of the induction statement for $q = p$. In order to do so we will proceed by induction on $n \in [l_p^+ + 1, \nu(p+1)]$. The next step starts the induction, by showing Part II for $n = l_p^+ + 1$.

Step 4: $\xi_{l_p^+ +1}(\beta, l_p^-) \in \overline{B_{\frac{1}{\alpha}}(0)}$ *implies* $\xi_j(\beta, l_p^-) \in \overline{B_{\frac{1}{\alpha}}(0)}\ \forall j \in R_{l_p^+ +1}$

Assume that $\xi_{l_p^+ +1}(\beta, l_p^-) \in \overline{B_{\frac{1}{\alpha}}(0)}$. As we are in the case of one-sided forcing, $\xi_j(\beta, l_p^-) \leq -\frac{1}{\alpha}$ for any $j \in [1, l_p^+]$ implies $\xi_{l_p^+ +1}(\beta, l_p^-) \leq -\frac{1}{\alpha}$ (compare the discussion below (6.12)). Therefore, it suffices to show that for any $j \in R_{l_p^+ +1} \setminus \{l_p^+ + 1\}$

$$\xi_j(\beta, l_p^-) \geq \frac{1}{\alpha} \quad \textit{implies} \quad \xi_{l_p^+ +1}(\beta, l_p^-) \geq \frac{1}{\alpha} . \tag{6.13}$$

Using the two claims below, this can be done as follows: Suppose $j \in R_{l_p^+ +1}$ and $\xi_j(\beta, l_p^-) \geq \frac{1}{\alpha}$. Then $d(\omega_j, 0) \geq \frac{3\gamma}{L_2}$ by Lemma 5.26(b), such that Lemma 4.3 implies $\xi_{j+1}(\beta, l_p^-) \geq \gamma \geq \frac{2}{\alpha}$. Therefore (6.13) follows directly from Claim 6.5 with $k = j + 1$, provided $j + 1 \in R_{l_p^+ +1}$. On the other hand, if $j + 1 \in \Gamma_{l_p^+ +1}$ then Claim 6.6 (with $k = j$) yields the existence of a suitable $\tilde{k}$, such that (6.13) follows again from Claim 6.5. As $R_{l_p^+ +1} \cup \Gamma_{l_p^+ +1} = [1, l_p^+ + 1]$, this covers all possible cases.

CLAIM 6.5. *Suppose* $\xi_k(\beta, l_p^-) \geq \frac{2}{\alpha}$ *for some* $k \in R_{l_p^+ +1}$. *Then* $\xi_{l_p^+ +1}(\beta, l_p^-) > \frac{1}{\alpha}$.

PROOF. Let $\beta^* := \beta^+_{p-1,\nu(p)}$ as in Step 3. Note that $\xi_{l_p^+ +1}(\beta, l_p^-) \in \overline{B_{\frac{1}{\alpha}}(0)}$ implies $|\beta - \beta^*| \leq \alpha^{-p}$ by Step 3. Further, we can again apply Part II of the

---

[7]Again, the assumptions of the lemma with $\epsilon = \alpha^{-p}$ are all satisfied: We have err$(\dots) \leq K \cdot \epsilon$ as before. (6.8) together with Lemma 5.27(a) implies

$$\tau(n) \leq \#([1, l_p^+] \setminus R_{\nu(p)}) \leq \frac{2p-3}{4} .$$

and similarly $\tau(j) \leq \frac{j}{8}$). As $l_p^+ + 1 \in R_{\nu(p)}$ by (5.3) we also have $x_{n+1}^1 = \xi_{l_p^+ +1}(\beta^*, l_{p-1}^-) \in \overline{B_{\frac{1}{\alpha}}(0)}$. Further $\tau(n) - \tau(j) \leq \frac{n-j}{6}$ follows from (6.7), and $n = l_p^+ \geq 5p$ by (6.6).

induction statement to $q = p-1$ and $n = \nu(p)$. As $R_{l_p^+ +1} \subseteq R_{\nu(p)}$ (see (5.5)) we obtain

$$\xi_j(\beta^*, l_{p-1}^-) \in \overline{B_{\frac{1}{\alpha}}(0)} \quad \forall j \in R_{l_p^+ +1} . \tag{6.14}$$

The claim now follows from Lemma 5.6(a), which we apply to compare the orbits[8]

$$x_1^1, \dots, x_n^1 := \xi_k(\beta^*, l_{p-1}^-), \dots, \xi_{l_p^+}(\beta^*, l_{p-1}^-) \tag{6.15}$$

and

$$x_1^2, \dots, x_n^2 := \xi_k(\beta, l_p^-), \dots, \xi_{l_p^+}(\beta, l_p^-) . \tag{6.16}$$

Thus we obtain $\xi_{l_p^+ +1}(\beta, l_p^-) = x_{n+1}^2 \geq \frac{2}{\alpha}$.

□

CLAIM 6.6. *Suppose $k \in R_{l_p^+ +1}$, $k+1 \in \Gamma_{l_p^+ +1}$ and $\xi_k(\beta, l_p^-) \geq \frac{1}{\alpha}$. Then there exists some $\tilde{k} \in R_{l_p^+ +1}$ with $\xi_{\tilde{k}}(\beta, l_p^-) \geq \frac{2}{\alpha}$.*

PROOF. First of all, note that $\Gamma_{l_p^+ +1} = \Gamma_{l_p^+}$ by (5.10) and $\Gamma_{l_p^+} = \bigcup_{J \in \mathcal{J}_{l_p^+}} \Gamma(J)$ by (5.7). Therefore, there must be some $J_1 \in \mathcal{J}_{l_p^+}$ such that $k+1 \in \Gamma(J_1)$. Let $m_1 := m_{J_1}$ and $p_1 := p(m_1)$. As $\Gamma(J_1) = J_1^- \cup \Gamma^+(J_1)$, we have two possibilities:

Either $k+1 \in J_1^-$, which means that $j+1 = \lambda^-(m_1)$ (as $J_1^- = [\lambda^-(m_1), m_1]$ is an interval and we assumed $k \in R_{l_p^+ +1}$). In this case define $m = m_1$ and $t = 0$.

The other alternative is that $k+1 \in \Gamma^+(J_1)$, and in this case we have to '*go backwards through the recursive structure of the set $R_{l_p^+ +1}$*', until we arrive at the first alternative:

As $\Gamma^+(J_1) = \Gamma_{l_{p_1}^+} + m_1$ by (5.8), $k+1 \in \Gamma^+(J_1)$ means that $k - m_1 + 1 \in \Gamma_{l_{p_1}^+}$. Hence, similar to before there exists some $J_2 \in \mathcal{J}_{l_{p_1}^+}$ such that either $k - m_1 + 1 = \lambda^-(m_{J_2})$ or $k - m_1 + 1 \in \Gamma^+(J_2)$. Let $m_2 := m_{J_2}$ and $p_2 := p(m_2)$. If we are in the second case where $j - m_1 + 1 \in \Gamma^+(J_2)$ we continue like this, but after finitely many steps the procedure will stop and we arrive at the first alternative. This is true because in each step the p_i become smaller, more precisely $p_{i+1} \leq p_i - 3$,[9] and finally $\Gamma_{l_1^+}$ is empty. Thus we obtain two sequences $p_1 > \ldots > p_r \geq 0$ and $m_1 > \ldots > m_r$ with $p_i = p(m_i)$, such that $k - \sum_{i=1}^{r-1} m_i + 1 = \lambda^-(m_r)$ for some $r \in \mathbb{N}$. Let $m := m_r$ and $t := \sum_{i=1}^{r-1} m_i$, such that $p_r = p(m)$ and $k+1-t = \lambda^-(m)$. Note that for $r = 1$ this coincides with the above definitions of m and t in the first case. We have

$$\begin{aligned} d(\omega_t, 0) \ &\leq \ \sum_{i=1}^{r-1} d(\omega_{m_i}, 0) \ \leq \ \sum_{i=1}^{r-1} \frac{\alpha^{-(p_i-1)}}{L_2} \\ &\leq \ \frac{\alpha^{-(p(m)+2)} \cdot S_\infty(\alpha)}{L_2} \ \overset{(6.5)}{\leq} \ \frac{1}{4} \cdot \frac{\alpha^{-(p(m)+1)}}{L_2} . \end{aligned} \tag{6.17}$$

---

[8]We choose $\epsilon = \alpha^{-p}$. $\mathrm{err}(\ldots) \leq K \cdot \epsilon$ follows from $|\beta - \beta^*| \leq \alpha^{-p}$. As $k \in R_{l_p^+ +1}$ by assumption and $l_p^+ + 1 \in R_{l_p^+ +1}$ by (5.3), we have $x_1^1, x_{n+1}^1 \in \overline{B_{\frac{1}{\alpha}}(0)}$ by (6.14). Finally $\tau(n) \leq \min\{0, \frac{2p-3}{4}\}$ and $\tau(n) - \tau(j) \leq \frac{n-j}{6}$ follow from Lemma 5.27(a) and (6.7).

[9]Note that there is no $J \in \mathcal{J}_{l_{p_i}^+}$ with $p_J \geq p_i - 2$ by (6.6).

Now choose some $q' \geq p(m) \geq 1$ such that $l^+_{q'} + 1 \leq m \leq \nu(q'+1)$. This is possible as $m \geq \nu(p(m)) \geq l^+_{p(m)} + 1$, and because the intervals $[l^+_q + 1, \nu(q)]$ overlap by (6.6). In addition, we can choose $q' < p-1$ as $m \leq l^+_p + 1 < \nu(p-1)$.

We now want to apply Lemma 6.3 with $\beta^* := \beta^+_{q',m}$ $q = p(m)$, $l^* = l^-_{q'}$ $l = l^-_p$ and $k = t$. Note that we can apply Part II of the induction statement with $q = q'$ and $n = m$ to obtain that $\beta^* \in \left[1 + \frac{1}{\sqrt{\alpha}}, 1 + \frac{3}{\sqrt{\alpha}}\right]$,

$$\xi_j(\beta^*, l^-_{q'}) \geq \gamma \quad \forall j \in [l^-_{q'}, 0] \setminus \Omega_\infty \tag{6.18}$$

and

$$\xi_j(\beta^*, l^-_{q'}) \in \overline{B_{\frac{1}{\alpha}}(0)} \quad \forall j \in R_m . \tag{6.19}$$

Further, Part III of the induction statement[10] together with Step 3 imply that

$$|\beta - \beta^*| \leq |\beta - \beta_{p-1,\nu(p)}| + |\beta_{p-1,\nu(p)} - \beta^*| \leq \alpha^{-p} + \alpha^{-q'} \leq 2\alpha^{-q'} \tag{6.20}$$

and finally $\xi_{k+1}(\beta, l^-_p) \geq \gamma$ if $\xi_k(\beta, l^-_p) \geq \frac{1}{\alpha}$ by Lemma 4.3 . Thus Lemma 6.3 yields

$$\{j \in [\lambda^-(m), m] \mid \xi_{j+t}(\beta, l^-_p) < \gamma\} \subseteq \tilde{\Omega}_{p(m)-2} . \tag{6.21}$$

Consequently (Lemma 5.10(b))

$$\{j \in [-l^-_{p(m)}, 0] \mid \xi_{j+m+t}(\beta, l^-_p) < \gamma\} \subseteq \Omega_\infty . \tag{6.22}$$

This means that we can compare the two sequences

$$x^1_1, \ldots, x^1_n := \xi_{-l^-_{p(m)}}(\beta^*, l^-_{q'}), \ldots, \xi_{-1}(\beta^*, l^-_{q'}) \tag{6.23}$$

and

$$x^2_1, \ldots, x^2_n := \xi_{m+t-l^-_{p(m)}}(\beta, l^-_p), \ldots, \xi_{m+t-1}(\beta, l^-_p) \tag{6.24}$$

via Lemma 5.2 with $\epsilon := L_2 \cdot d(\omega_m, 0) \in (\alpha^{-p(m)}, \alpha^{-(p(m)-1)}]$ to obtain that[11]

$$|\xi_{m+t}(\beta, l^-_p) - \xi_0(\beta^*, l^-_{q'})| \leq \epsilon \cdot (6 + K \cdot S_\infty(\alpha^{\frac{1}{4}})) . \tag{6.25}$$

As $d(\omega_{m+t}, 0) \geq \frac{3}{4} \cdot \frac{\epsilon}{L_2}$ (see (6.17)), it follows from (2.7) and (2.11) that

$$\begin{aligned} &\xi_{m+t+1}(\beta, l^-_p) \geq \\ &\geq \xi_1(\beta^*, l^-_{q'}) + \frac{3\epsilon}{4} - \epsilon \cdot \frac{6 + K \cdot S_\infty(\alpha^{\frac{1}{4}})}{2\sqrt{\alpha}} \overset{(6.4)}{\geq} \xi_1(\beta^*, l^-_{q'}) + \frac{\epsilon}{2} . \end{aligned} \tag{6.26}$$

Now first assume $p(m) \geq 2$, such that $\epsilon \leq \frac{1}{\alpha}$. (The case $p(m) = 1$ has to be treated separately, see below.) Then we can apply Lemma 5.6(b) to compare the orbits

$$x^1_1, \ldots, x^1_n := \xi_1(\beta^*, l^-_{q'}), \ldots, \xi_{l^+_{p(m)}}(\beta^*, l^-_{q'}) \tag{6.27}$$

and

$$x^2_1, \ldots, x^2_n := \xi_{m+t+1}(\beta, l^-_p), \ldots, \xi_{m+t+l^+_{p(m)}}(\beta, l^-_p) \tag{6.28}$$

---

[10] With $q_1 = q'$, $q = p-1$, $n_1 = m$ and $n_2 = \nu(p-1)$.

[11] We have $q = p(m) - 1$. Note that $d(\omega_{m+t}, 0) \leq \frac{2\epsilon}{L_2}$ (see (6.17)) and $|\beta - \beta^*| \leq 2\alpha^{-q'} \leq 2\epsilon$ by (6.20), such that $\text{err}(\ldots) \leq K \cdot \epsilon$ by Remark 4.5 . Further, it follows from (6.18) and (6.22) that $\eta(j,n) \leq \#([-(n-j), -1] \cap \Omega_\infty) \leq \frac{n-j}{10}$ (see (5.2)). Finally $n = l^-_{p(m)} \geq 4p(m)$ by (6.6), such that $\alpha^{-\frac{1}{4}n} \leq \epsilon$.

to conclude that[12]

$$\xi_{m+t+l_q^++1}(\beta, l_p^-) \geq \frac{2}{\alpha}. \tag{6.29}$$

As $J_r = J(m)$ is a maximal interval in $\Gamma_{l^+_{p_{r-1}}}$ we have $\lambda^+(m)+1 \in R_{l^+_{p_{r-1}}}$. Therefore $\lambda^+(m)+1+t \in R_{l_p^++1}$ follows from the recursive structure of this set. Consequently, we can choose $\tilde{k} = m+t+l^+_{p(m)}+1$.

Finally, suppose $p(m) = 1$. In this case we still have $\xi_{m+t+1}(\beta, l_p^-) \geq \xi_1(\beta^*, l_{q'}^-) + \frac{\epsilon}{2}$ by (6.26). There are two possibilities: Either $\xi_{m+t+1}(\beta, l_p^-) \geq \frac{2}{\alpha}$. As $m+1 \in R_{l_{p_{r-1}}}$ (see Remark 5.23(c)), we have $m+t+1 \in R_{l_p^++1}$ due to the recursive structure of this set. Thus, we can choose $\tilde{k} = m+t+1$. On the other hand, if $\xi_{m+t+1}(\beta, l_p^-) \in B_{\frac{2}{\alpha}}(0)$ then we can apply (2.7) again and obtain

$$\xi_{m+t+2}(\beta, l_p^-) \geq \xi_2(\beta^*, l_{q'}^-) + 2\sqrt{\alpha}\cdot\epsilon - K\epsilon \overset{(5.1)}{\geq} \xi_2(\beta^*, l_{q'}^-) + \sqrt{\alpha}\cdot\epsilon \geq \frac{1}{\alpha}$$

by (2.2), as $\xi_2(\beta^*, l_q^-) \in \overline{B_{\frac{1}{\alpha}}(0)}$ by (6.19) and $\epsilon \geq \frac{1}{\alpha}$. Thus, we can choose $\tilde{k} = m+t+2$ in this case. Note that $m+t+2$ is contained in $R_{l_p^++1}$ for the same reasons as $m+t+1$.

□

■

Now we can show that $\xi_{l_p^++1}(\beta, l_p^-) \in \overline{B_{\frac{1}{\alpha}}(0)}$ implies $\beta \in \left[1+\frac{1}{\sqrt{\alpha}}, 1+\frac{3}{\sqrt{\alpha}}\right]$ and thus complete the proof of Part I of the induction statement: Suppose $\xi_{l_p^++1}(\beta, l_p^-) \in \overline{B_{\frac{1}{\alpha}}(0)}$. By Steps 2 and 3 we know that $\xi_0(\beta, l_p^-) \geq \gamma$. This implies that

$$\xi_1(\beta, l_p^-) \in \left[1+\frac{3}{2\sqrt{\alpha}} - \beta, 1+\frac{3}{\sqrt{\alpha}} - \frac{1}{\alpha} - \beta\right]$$

(see assumptions (2.4) and (2.7)). As Step 4 implies that $\xi_1(\beta, l_p^-) \in \overline{B_{\frac{1}{\alpha}}(0)}$, and $\frac{1}{2\sqrt{\alpha}} \geq \frac{1}{\alpha}$ by (5.1), this gives the required estimate.

Step 5: *Part II of the induction statement implies Part III*

Actually, the situation is a little bit more complicated than the headline above may suggest. In fact, the both remaining parts of the induction statement have to be proved simultaneously by induction on n. However, in each step of the induction Part II will imply Part III.

---

[12]We have $q = p(m) - 1$ and $\mathrm{err}(\ldots) \leq K \cdot \epsilon$ as before (see Footnote 11). (6.19) yields that $x^1_{n+1} \in \overline{B_{\frac{1}{\alpha}}(0)}$ (note that $l_q^+ + 1 \in R_m$ by (5.3)). Further, we have

$$\tau(n) \leq \#\left([1, l_q^+] \setminus R_m\right) \leq \max\left\{0, \frac{2q-3}{4}\right\}$$

and $\tau(j) \leq \#([1,j] \setminus R_m) \leq \frac{j}{8}$ by Lemma 5.27(a). $\tau(n) - \tau(j) \leq \frac{n-j}{6}$ follows again from (6.7), and finally $n = l^+_{p(m)} \geq 5(p(m)-1)$.

In order to make this more precise, assume that Part II with $q = p$ holds for all $n \leq N$, with $N \in [l_p^+ + 1, \nu(p+1)]$. What we will now show is that in this case Part III(a) holds as well whenever $n_1, n_2 \leq N$, and similarly Part III(b) holds whenever $n_2 \leq N$.

Suppose that $N \in [l_p^+ + 1, \nu(p+1)]$ and Part II with $q = p$ holds for all $n \leq N$. Further, let $n_2 \leq N$ and $n_1 \in R_{n_2}$. Then the Part II of the induction statement applied to $q = p$ and $n = n_2$ yields that $\xi_{n_1}(\beta^+_{q,n_2}, l_p^-) \in \overline{B_{\frac{1}{\alpha}}(0)}$, and for $n = n_1$ we obtain that $\xi_{n_1}(\beta, l_p^-) \in \overline{B_{\frac{1}{\alpha}}(0)}$ implies $\xi_j(\beta, l_p^-) \in \overline{B_{\frac{1}{\alpha}}(0)}$ $\forall j \in R_{n_1}$. Thus all the assumption of Lemma 6.2 (with $n = n_1$, $\beta_1 = \beta^+_{p,n_1}$ and $\beta_2 = \beta^+_{p,n_2}$) are satisfied, such that

$$|\beta^+_{p,n_1} - \beta^+_{p,n_2}| \leq 2\alpha^{-\frac{n_1}{4}}$$

as required.

For Part III(b) let $q_1 < p$, $n_1 \in [l_{q_1}^+ + 1, \nu(q_1 + 1)]$ and $n_2 \in [l_p^+ + 1, N]$. First suppose $q_1 < p - 1$. Then Part III(b) of the induction statement (with $q = p - 1$ and $n_2 = \nu(p)$) yields

$$|\beta^+_{q_1,n_1} - \beta^+_{p-1,\nu(p)}| \leq 3 \cdot \sum_{i=q_1+1}^{p-1} \alpha^{-i} .$$

Further, Part II of the induction statement (with $q = p$ and $n = n_2$) yields that $\xi_{l_p^+ + 1}(\beta^+_{p,n_2}, l_p^-) \in \overline{B_{\frac{1}{\alpha}}(0)}$ (note that $l_p^+ + 1 \in R_{n_2}$ by (5.3)), and consequently

$$|\beta^+_{p-1,\nu(p)} - \beta^+_{p,n_2}| \leq \alpha^{-p}$$

by Step III. Altogether, we obtain

$$|\beta^+_{q_1,n_1} - \beta^+_{p,n_2}| \leq |\beta^+_{q_1,n_1} - \beta^+_{p-1,\nu(p)}| + |\beta^+_{p-1,\nu(p)} - \beta^+_{p,n_2}| \leq 3 \cdot \sum_{i=q_1+1}^{p} \alpha^{-i} .$$

On the other hand, if $q_1 = p-1$ then Part III(a) (with $q = q_1 = p-1$ and $n_2 = \nu(p)$) in combination with $n_1 \geq l_{p-1}^+ \geq 4p$ (see (6.6)) yields

$$|\beta^+_{q_1,n_1} - \beta^+_{p-1,\nu(p)}| \leq 2\alpha^{-fracn_1 4} \leq 2\alpha^{-p} ,$$

such that

$$|\beta^+_{q_1,n_1} - \beta^+_{p,n_2}| \leq |\beta^+_{q_1,n_1} - \beta^+_{p-1,\nu(p)}| + |\beta^+_{p-1,\nu(p)} - \beta^+_{p,n_2}| \leq 3\alpha^{-p}$$

as required. Finally, note that

$$3 \cdot \sum_{i=q_1+1}^{p} \alpha^{-i} \leq \frac{3S_\infty(\alpha)}{\alpha} \cdot \alpha^{-q_1} \leq \alpha^{-q_1}$$

by (6.5).

■

Now we can already use the parameter estimates up to N (in the way mentioned above) during the induction step $N \to N + 1$ in the proof of Part II.

Step 6: *Proof of Part II for $q = p$.*

In order to prove Part II of the induction statement for $q = p$, we will proceed by induction on n. Steps 2–4 show that the statement holds for $n = l_p^+ + 1$. Suppose now that it holds for all $n \leq N$, where $N \in [l_p^+ + 1, \nu(p+1) - 1]$. We have to show that it then holds for $N+1$ as well. In order to do so, we distinguish three different cases: First, if $N+1$ is not admissible there is nothing to prove. Secondly, if both N and $N+1$ are admissible then necessarily $p(N) = 0$, otherwise $N+1$ would be contained in $J(N)$. Thus $d(\omega_N, 0) \geq \frac{4\gamma}{L_2}$, and in addition Part II of the induction statement with $q = p$ and $n = N$ implies that $\beta_{p,N}^{\pm} \in \left[1 + \frac{1}{\sqrt{\alpha}}, 1 + \frac{3}{\sqrt{\alpha}}\right]$. Therefore Lemma 4.3 yields that

$$\xi_{N+1}(\beta_{p,N}^+, l_p^-) > \frac{1}{\alpha}$$

and

$$\xi_{N+1}(\beta_{p,N}^-, l_p^-) < -\frac{1}{\alpha} .$$

Consequently $\xi_{N+1}(\beta, l_p^-) \in \overline{B_{\frac{1}{\alpha}}(0)}$ implies that $\xi_N(\beta, l_p^-) \in \overline{B_{\frac{1}{\alpha}}(0)}$, and everything else follows from Part II of the induction statement for $n = N$.

Thus, it remains to treat the case where $N+1$ is admissible but $N \notin A_N$. By (5.12) this also means that $N \notin A_{N+1}$. Consequently there exists an interval $J \in \mathcal{J}_{N+1}$ which contains N, such that $J = [t, N]$ where $t := \lambda^-(m_J)$. Note that $t-1, t, m_J \in A_{m_J}$ by Lemma 5.20(a). In particular m_J and $t-1$ are admissible. First of all, we will prove the following claim.

CLAIM 6.7. *$\xi_{N+1}(\beta, l_p^-) \in \overline{B_{\frac{1}{\alpha}}(0)}$ implies $\xi_{t-1}(\beta, l_p^-) \in \overline{B_{\frac{1}{\alpha}}(0)}$.*

PROOF. It suffices to show that

$$\xi_{N+1}(\beta_{p,t-1}^+, l_p^-) > \frac{1}{\alpha} \tag{6.30}$$

(see (6.8)–(6.12)). Let $m := m_J$, $\beta^+ := \beta_{p,t-1}^+$ and $\beta^* := \beta_{p,m}^+$. Using Part II of the induction statement (with $q = p$ and $n = m$) we obtain

$$\xi_j(\beta^*, l_p^-) \in \overline{B_{\frac{1}{\alpha}}(0)} \quad \forall j \in R_m \tag{6.31}$$

and $\beta^* \in \left[1 + \frac{1}{\sqrt{\alpha}}, 1 + \frac{3}{\sqrt{\alpha}}\right]$. Further, the same statement with $n = t-1$ implies

$$\xi_j(\beta^+, l_p^-) \geq \gamma \quad \forall j \in [-l_p^-, 0] \setminus \Omega_\infty \tag{6.32}$$

and

$$\xi_j(\beta^+, l_p^-) \in \overline{B_{\frac{1}{\alpha}}(0)} \quad \forall j \in R_{t-1} . \tag{6.33}$$

Finally Part III(a) of the induction statement (with $q = p$, $n_1 = t-1$ and $n_2 = m$) yields that

$$|\beta^+ - \beta^*| \leq \alpha^{-\frac{(t-1)}{4}} \leq \alpha^{-\frac{l_p^+}{4}} \overset{(6.6)}{\leq} \alpha^{-(p+1)} . \tag{6.34}$$

Note that $t-1$ is contained in $\Omega_\infty(m)$ by (5.8), and as $l_p^+ + 1 \notin \Omega_\infty$ this interval must be to the right of $l_p^+ + 1$. Therefore $t - 1 > l_p^+ + 1$. Now all the assumptions for the application of Lemma 6.3 are satisfied[13] and we obtain

$$\{j \in [t, m] \mid \xi_j(\beta^+, l_p^-) < \gamma\} \subseteq \tilde{\Omega}_{p(m)-2} \, . \tag{6.35}$$

Using Lemma 5.10(b) this further means that

$$\{j \in [-l_{p(m)}^-, 0] \mid \xi_{j+m}(\beta^+, l_p^-) < \gamma\} \subseteq \Omega_\infty \, . \tag{6.36}$$

Now we compare the orbits

$$x_1^1, \ldots, x_n^1 := \xi_{-l_{p(m)}^-}(\beta^+, l_p^-), \ldots, \xi_{-1}(\beta^+, l_p^-) \tag{6.37}$$

and

$$x_1^2, \ldots, x_n^2 := \xi_t(\beta^+, l_p^-), \ldots, \xi_{m-1}(\beta^+, l_p^-) \, , \tag{6.38}$$

using Lemma 5.2 with $\epsilon := L_2 \cdot d(\omega_m, 0) \in [\alpha^{-p(m)}, \alpha^{-(p(m)-1)})$, to conclude that[14]

$$|\xi_m(\beta^+, l_p^-) - \xi_0(\beta^+, l_p^-)| \leq \epsilon \cdot (6 + K \cdot S_\infty(\alpha^{\frac{1}{4}})) \, .$$

As (6.32) and (6.36) in particular imply that $\xi_m(\beta^+, l_p^-), \xi_0(\beta^+, l_p^-) \geq \gamma$, it follows from (2.7) and (2.11) that

$$\xi_{m+1}(\beta^+, l_p^-) \geq$$

$$\geq \xi_1(\beta^+, l_p^-) + \epsilon - \frac{\epsilon \cdot (6 + K \cdot S_\infty(\alpha^{\frac{1}{4}}))}{2\sqrt{\alpha}} \stackrel{(6.4)}{\geq} \xi_1(\beta^+, l_p^-) + \frac{\epsilon}{2} \, . \tag{6.39}$$

Now first assume $p(m) \geq 2$, such that $d(\omega_m, 0) \leq \frac{\alpha^{-1}}{L_2}$. Then we can apply Lemma 5.6(b)[15] to the sequences

$$x_1^1, \ldots, x_n^1 := \xi_1(\beta^+, l_p^-), \ldots, \xi_{l_{p(m)}^+}(\beta^+, l_p^-) \tag{6.40}$$

and

$$x_1^2, \ldots, x_n^2 := \xi_{m+1}(\beta^+, l_p^-), \ldots, \xi_N(\beta^+, l_p^-) \, , \tag{6.41}$$

which yields that $\xi_{N+1}(\beta^+, l_p^-) = x_{n+1}^2 \geq \frac{2}{\alpha}$ as required for (6.30).

It remains to address the case $p(m) = 1$. Note that in this case $p(j) = 0 \; \forall j \in [m+1, N]$ (see Lemma 5.20(a)). There are two possibilities: Either $\xi_{m+1}(\beta^+, l_p^-) \geq \frac{1}{\alpha}$, in which case $\xi_{N+1}(\beta^+, l_p^-) \geq \gamma > \frac{1}{\alpha}$ follows from the repeated application of Lemma 4.3 . Otherwise, $\xi_{m+1}(\beta^+, l_p^-) \in \overline{B_{\frac{1}{\alpha}}(0)}$. As $1, 2 \in R_{t-1}$ (see (5.3)) it

---

[13] With β^* and m as above, $q = p(m)$ $(\leq p)$, $l = l^* = l_p^-$, $k = 0$ and $\beta = \beta^+$. Note that $p(t-1) = 0$ as $t-1$ is admissible, and $\xi_{t-1}(\beta^+, l_p^-) = \frac{1}{\alpha}$ by definition of $\beta^+ = \beta_{p,t-1}^+$. Therefore Lemma 4.3 implies $\xi_{m-l_{p(m)}^-}(\beta^+, l_p^-) = \xi_t(\beta^+, l_p^-) \geq \gamma$.

[14] As $\beta_1 = \beta_2 = \beta^+$ we have $\mathrm{err}(\ldots) \leq K\epsilon$, see Remark 4.5. (6.32) and (6.36) imply that

$$\eta(j, n) \leq \#([-(n-j), -1] \cap \Omega_\infty) \leq \frac{n-j}{10}$$

by (5.2). Finally $n = l_{p(m)}^- \geq 4(p(m) + 1)$ by (6.6), such that $\alpha^{-\frac{n}{4}} \leq \alpha^{-(p(m)+1)} \leq \epsilon$.

[15] With $\epsilon = L_2 \cdot d(\omega_m, 0)$ as above, such that $q = p(m) - 1$. $\mathrm{err}(\ldots) \leq K \cdot \epsilon$ follows again from Remark 4.5. $x_{n+1}^1 \in \overline{B_{\frac{1}{\alpha}}(0)}$ follows from (6.33) as $l_{p(m)}^+ + 1 \in R_{t-1}$ by (5.3). (6.33) also implies $\tau(n) \leq \frac{2p(m)-5}{4}$ and $\tau(j) \leq \frac{j}{8} \; \forall j \in [1, n]$ by Lemma 5.27(a) and $\tau(n) - \tau(j) \leq \frac{n-j}{6}$ by (6.7). Finally $n = l_{p(m)}^+ \geq 5(p(m) - 1)$ by (6.6).

follows from (6.33) that $\xi_1(\beta^+, l_p^-), \xi_2(\beta^+, l_p^-) \in \overline{B_{\frac{1}{\alpha}}(0)}$ as well. Therefore (2.6) implies that

$$\begin{aligned} \xi_{m+2}(\beta^+, l_p^-) &\geq \xi_2(\beta^+, l_p^-) + 2\sqrt{\alpha}\cdot\epsilon - K\cdot\epsilon \\ &\overset{(5.1)}{\geq} \xi_2(\beta^+, l_p^-) + \sqrt{\alpha}\cdot\epsilon \geq \frac{2}{\alpha} \end{aligned} \tag{6.42}$$

as $\epsilon \geq \frac{1}{\alpha}$ in this case. Again, we obtain $\xi_{N+1}(\beta^+, l_p^-) \geq \gamma > \frac{1}{\alpha}$ by repeated application of Lemma 4.3 .

□

Now suppose $\xi_{N+1}(\beta, l_p^-) \in \overline{B_{\frac{1}{\alpha}}(0)}$. Then by Claim 6.7 there holds $\xi_{t-1}(\beta, l_p^-) \in \overline{B_{\frac{1}{\alpha}}(0)}$. As we can already apply Part II of the induction statement with $q = p$ and $n = t-1$, this further implies (6.13), $\beta \in \left[1 + \frac{1}{\sqrt{\alpha}}, 1 + \frac{3}{\sqrt{\alpha}}\right]$ and

$$\xi_j(\beta, l_p^-) \in \overline{B_{\frac{1}{\alpha}}(0)} \quad \forall j \in R_{t-1} .$$

Note that $R_{N+1} \cap [1, t-1] = R_{t-1}$ (see Remark 5.23(b)), such that $R_{N+1} = R_{t-1} \cup R(J) \cup \{N+1\}$. Therefore, in order to complete this step and thereby the proof of Induction scheme 6.1, it only remains to show that

CLAIM 6.8. *$\xi_{N+1}(\beta, l_p^-) \in \overline{B_{\frac{1}{\alpha}}(0)}$ implies $\xi_j(\beta, l_p^-) \in \overline{B_{\frac{1}{\alpha}}(0)}$ $\forall j \in R(J)$.*

PROOF. The proof of this statement is very similar to the proof of Step 4, and likewise we will use two further claims, namely Claim 6.9 and Claim 6.10 below, which are the analogues of Claim 6.5 and Claim 6.6 . Suppose $\xi_j(\beta, l_p^-) > \frac{1}{\alpha}$ for some $j \in R(J)$. We have to distinguish two cases (note that $R(J) \cup \Gamma^+(J) = J^+$): Either $j+1 \in R(J)$. As $d(\omega_j, 0) \geq \frac{3\gamma}{L_2}$ by Lemma 5.26(b), Lemma 4.3 implies $\xi_{j+1}(\beta, l_p^-) \geq \gamma \geq \frac{2}{\alpha}$. Therefore we can apply Claim 6.9 with $k = j+1$. On the other hand, if $j+1 \notin R(J)$, then Claim 6.10 (with $k = j$) yields the existence of a suitable $\tilde{k}$ and we can again apply Claim 6.9, this time with $k = \tilde{k}$. In both cases we obtain that $\xi_j(\beta, l_p^-) \geq \frac{1}{\alpha}$ implies $\xi_{N+1}(\beta, l_p^-) > \frac{1}{\alpha}$. As we are in the case of one-sided forcing, the fact that $\xi_j(\beta, l_p^-) \leq -\frac{1}{\alpha}$ implies $\xi_{N+1}(\beta, l_p^-) < -\frac{1}{\alpha}$ is obvious. This proves the claim.

□

CLAIM 6.9. *Suppose $\xi_k(\beta, l_p^-) \geq \frac{2}{\alpha}$ for some $k \in R(J)$. Then $\xi_{N+1}(\beta, l_p^-) > \frac{1}{\alpha}$.*

PROOF. First of all, if $p_J = 1$ then $p(j) = 0$ $\forall j \in J^+$ by Lemma 5.20(a), and the claim follows from the repeated application of Lemma 4.3 . Thus we can assume $p_J \geq 2$. Claim 6.7 together with Part II of the induction statement with $q = p$ and $n = t-1$ imply that

$$\xi_j(\beta, l_p^-) \in \overline{B_{\frac{1}{\alpha}}(0)} \quad \forall j \in R_{t-1} \supseteq R_{l^+_{p_J}+1} \supseteq R_{l^+_{p_J}} \tag{6.43}$$

(see (5.5) for the inclusions). Consequently, we can apply Lemma 5.6(a)[16] to the sequences

$$x_1^1, \ldots, x_n^1 \; := \; \xi_{k-m_J}(\beta, l_p^-), \ldots, \xi_{l_{p_J}^+}(\beta, l_p^-) \tag{6.44}$$

and

$$x_1^2, \ldots, x_n^2 \; := \; \xi_k(\beta, l_p^-), \ldots, \xi_N(\beta, l_p^-) \; . \tag{6.45}$$

to obtain that $\xi_{N+1}(\beta, l_p^-) = x_{n+1}^2 \geq \frac{2}{\alpha}$.

□

CLAIM 6.10. *Suppose $k \in R(J)$, $k+1 \in \Gamma^+(J)$ and $\xi_k(\beta, l_p^-) \geq \frac{1}{\alpha}$. Then there exists some $\tilde{k} \in R(J)$ with $\xi_{\tilde{k}}(\beta, l_p^-) \geq \frac{2}{\alpha}$.*

PROOF. Let $J_1 := J$, $m_1 := m_J$ and $p_1 := p_J$. As in the proof of Claim 6.6 we can find sequences $p_1 > \ldots > p_r \geq 0$ and $m_1 > \ldots > m_r \in [1, l_{p_{r-1}}^+]$ with $p_i = p(m_i) \leq p_{i-1} - 3$, such that $k - \sum_{i=1}^{r-1} m_i + 1 = \lambda^-(m_r)$ for some $r \in \mathbb{N}$. Let $m := m_r$ and $t := \sum_{i=1}^{r-1} m_i$. (The only difference to Claim 6.6 is that $r = 1$ is not possible.) Likewise, we have

$$d(\omega_t, 0) \;\leq\; \frac{1}{4} \cdot \frac{\alpha^{-(p(m)+1)}}{L_2} \;\leq\; \frac{d(\omega_m, 0)}{4} \; . \tag{6.46}$$

Again, we choose some $q' \geq p(m)$ such that $l_{q'}^+ + 1 \leq m \leq \nu(q'+1)$. As $m \leq l_{p_{r-1}}^+ \leq l_p^+ < \nu(p-2)$ (see (6.6)) we can assume that $q' \leq p-2$.

We now want to apply Lemma 6.3 with $\beta^* := \beta_{q',m}^+$ $q = p(m)$, $l^* = l_{q'}^-$ $l = l_p^-$ and $k = t$. In order to check the assumptions, note that we can apply Part II of the induction statement (with $q = q'$ and $n = m$) to $\beta^* := \beta_{q',m}$ and obtain that $\beta^* \in \left[1 + \frac{1}{\sqrt{\alpha}}, 1 + \frac{3}{\sqrt{\alpha}}\right]$,

$$\xi_j(\beta^*, l_{q'}^-) \;\geq\; \gamma \qquad \forall j \in [l_{q'}^-, 0] \setminus \Omega_\infty \tag{6.47}$$

and

$$\xi_j(\beta^*, l_{q'}^-) \;\in\; \overline{B_{\frac{1}{\alpha}}(0)} \qquad \forall j \in R_m \setminus \Omega_\infty \; . \tag{6.48}$$

In addition, Step 3[17] together with Part III of the induction statement[18] imply that

$$|\beta - \beta^*| \;\leq\; |\beta - \beta_{p-1,\nu(p)}| + |\beta_{p-1,\nu(p)} - \beta^*| \;\leq\; \alpha^{-p} + \alpha^{-q'} \;\leq\; 2\alpha^{-p(m)} \tag{6.49}$$

Finally, $\xi_{k+1}(\beta, l_p^-) \geq \gamma$ if $\xi_k(\beta, l_p^-) \geq \frac{1}{\alpha}$ by Lemma 4.3[19] . Thus Lemma 6.3 yields

$$\{j \in [\lambda^-(m), m] \mid \xi_{j+t}(\beta, l_p^-) < \gamma\} \;\subseteq\; \tilde{\Omega}_{p(m)-2} \; . \tag{6.50}$$

Consequently (Lemma 5.10(b))

$$\{j \in [-l_{p(m)}^-, 0] \mid \xi_{j+m+t}(\beta, l_p^-) < \gamma\} \;\subseteq\; \Omega_\infty \; . \tag{6.51}$$

---

[16] We choose $\epsilon = L_2 \cdot d(\omega_{m_{p_J}}, 0) \in [\alpha^{-p_J}, \alpha^{-(p_J-1)}]$, such that $q = p_J - 1$ and $\mathrm{err}(\ldots) \leq K \cdot \epsilon$. Note that $k \in R(J)$ implies $k - m_J \in R_{l_{p_J}^+}$ by (5.1), and further $l_{p_J}^+ + 1 \in R_{l_{p_J}^+ + 1}$ by (5.3). Therefore $x_1^1, x_{n+1}^1 \in \overline{B_{\frac{1}{\alpha}}(0)}$ by (6.43). Finally $\tau(n) \leq \min\{0, \frac{2p(m)-5}{4}\}$ by Lemma 5.27(a) and $\tau(n) - \tau(j) \leq \frac{n-j}{6}$ by (6.7).

[17] Note that $\xi_{N+1}(\beta, l_p^-) \in \overline{B_{\frac{1}{\alpha}}(0)}$ implies $\xi_{t-1}(\beta, l_p^-) \in \overline{B_{\frac{1}{\alpha}}(0)}$ by Claim 6.7, which in turn implies $\xi_{l_p^+ + 1}(\beta, l_p^-) \in \overline{B_{\frac{1}{\alpha}}(0)}$ as $l_p^+ + 1 \in R_{t-1}$, see (5.3).

[18] With $q_1 = q'$, $q = p - 1$, $n_1 = m$ and $n_2 = \nu(p-1)$.

[19] Note that $k + 1 = \lambda^-(m)$ in the claim above corresponds to $m + k - l_{p(m)}^-$ in Lemma 6.3 .

This means that we can compare the two sequences

$$x_1^1, \ldots, x_n^1 \;:=\; \xi_{-l_p^-(m)}(\beta^*, l_{q'}^-), \ldots, \xi_{-1}(\beta^*, l_{q'}^-) \tag{6.52}$$

and

$$x_1^2, \ldots, x_n^2 \;:=\; \xi_{m+t-l_{p(m)}^-}(\beta, l_p^-), \ldots, \xi_{m+t-1}(\beta, l_p^-) \tag{6.53}$$

via Lemma 5.2 with $\epsilon \;:=\; L_2 \cdot d(\omega_m, 0) \;\in (\alpha^{-p(m)}, \alpha^{-(p(m)-1)}]$ to obtain that[20]

$$|\xi_{m+t}(\beta, l_p^-) - \xi_0(\beta^*, l_{q'}^-)| \;\leq\; \epsilon \cdot (6 + K \cdot S_\infty(\alpha^{\frac{1}{4}})) \,. \tag{6.54}$$

Note that (6.47) and (6.51) in particular imply that $\xi_0(\beta^*, l_{q'}^-) \geq \gamma$ and $\xi_{m+t}(\beta, l_p^-) \geq \gamma$. As $d(\omega_{m+t}, 0) \geq \frac{3}{4} \cdot \frac{\epsilon}{L_2}$ (see (6.46)), (2.11) in combination with (2.7) therefore implies

$$\begin{aligned} &\xi_{m+t+1}(\beta, l_p^-) \;\geq \\ &\geq\; \xi_1(\beta^*, l_{q'}^-) + \frac{3\epsilon}{4} - \epsilon \cdot \frac{6 + K \cdot S_\infty(\alpha^{\frac{1}{4}})}{2\sqrt{\alpha}} \;\overset{(6.4)}{\geq}\; \xi_1(\beta^*, l_{q'}^-) + \frac{\epsilon}{2} \,. \end{aligned} \tag{6.55}$$

Now first assume $p(m) \geq 2$, such that $\epsilon \leq \frac{\alpha^{-1}}{L_2}$. (The case $p(m) = 1$ has to be treated separately, see below.) Then we can apply Lemma 5.6(b), with ϵ as above, to compare the orbits

$$x_1^1, \ldots, x_n^1 \;:=\; \xi_1(\beta^*, l_{q'}^-), \ldots, \xi_{l_{p(m)}^+}(\beta^*, l_{q'}^-) \tag{6.56}$$

and

$$x_1^2, \ldots, x_n^2 \;:=\; \xi_{m+t+1}(\beta, l_p^-), \ldots, \xi_{m+t+l_{p(m)}^+}(\beta, l_p^-) \tag{6.57}$$

to conclude that[21]

$$\xi_{m+t+l_{p(m)}^+ + 1}(\beta, l_p^-) \;\geq\; \frac{2}{\alpha} \,. \tag{6.58}$$

As $J_r = J(m)$ is a maximal interval in $\Gamma_{l_{p_{r-1}}^+}$ we have $\lambda^+(m)+1 \in R_{l_{p_{r-1}}^+}$. Therefore $\lambda^+(m) + 1 + t \in R(J)$ follows from the recursive structure of the regular sets. Consequently, we can choose $\tilde{k} = \lambda^+(m) + 1 + t = m + l_{p(m)}^+ + t + 1$.

Finally, suppose $p(m) = 1$. In this case we still have $\xi_{m+t+1}(\beta, l_p^-) \geq \xi_1(\beta^*, l_{q'}^-) + \frac{\epsilon}{2}$ by (6.55). There are two possibilities: Either $\xi_{m+t+1}(\beta, l_p^-) \geq \frac{2}{\alpha}$. As $m+1 \in R_{l_{p_{r-1}}}$ (see Remark 5.23(c)), we have $m+t+1 \in R(J)$ due to the recursive structure of this

[20] Note that $d(\omega_{m+t}, 0) \leq \frac{2\epsilon}{L_2}$ (see (6.46)) and $|\beta - \beta^*| \leq 2\alpha^{-p(m)} \leq 2\epsilon$ by (6.49), such that err$(\ldots) \leq K \cdot \epsilon$ by Remark 4.5 . Further, it follows from (6.47) and (6.51) that $\eta(j, n) = \#([-(n-j), -1] \cap \Omega_\infty) \leq \frac{n-j}{10}$ (see (5.2)). Finally $n = l_{p(m)}^- \geq 4p(m)$ by (6.6), such that $\alpha^{-\frac{n}{4}} \leq \epsilon$.

[21] We have $q = p(m) - 1$ and err$(\ldots) \leq K\epsilon$ before (see Footnote 20). (6.48) yields that $x_{n+1}^1 \in \overline{B_{\frac{1}{\alpha}}(0)}$ (note that $l_q^+ + 1 \in R_m$ by (5.3)). Further we have

$$\tau(n) \;\leq\; \#([1, l_{p(m)}^+] \setminus R_m \;\leq\; \max\left\{0, \frac{2p(m) - 5}{4}\right\}$$

as well as $\tau(j) \leq \#([1, j] \setminus R_m) \leq \frac{j}{8}$ by Lemma 5.27(a). $\tau(n) - \tau(j) \leq \frac{n-j}{6}$ follows again from (6.7), and finally $n = l_{p(m)}^+ \geq 5(p(m) - 1)$ by (6.6).

set. Thus, we can choose $\tilde{k} = m+t+1$. On the other hand, if $\xi_{m+t+1}(\beta, l_p^-) \in B_{\frac{2}{\alpha}}(0)$ then we can apply (2.7) again and obtain

$$\xi_{m+t+2}(\beta, l_p^-) \geq \xi_2(\beta^*, l_{q'}^-) + 2\sqrt{\alpha}\cdot\epsilon - K\epsilon \overset{(5.1)}{\geq} \xi_2(\beta^*, l_{q'}^-) + \sqrt{\alpha}\cdot\epsilon \geq \frac{1}{\alpha}$$

by (2.2), as $\xi_1(\beta^*, l_{q'}^-), \xi_2(\beta^*, l_{q'}^-) \in \overline{B_{\frac{1}{\alpha}}(0)}$ by (6.48) and $\epsilon \geq \frac{1}{\alpha}$. Thus, we can choose $\tilde{k} = m+t+2$ in this case. Note that $m+t+2$ is contained in $R(J)$ for the same reasons as $m+t+1$.

□

■

CHAPTER 7

Construction of the sink-source-orbits: Symmetric forcing

For the symmetric setting, we will use two additional assumptions on the parameters, namely

$$\frac{4\gamma}{L_2} + \frac{S_\infty(\alpha)}{\alpha \cdot L_2} < \frac{1}{2}\,; \tag{7.1}$$

$$g_{|B_{\frac{4\gamma}{L_2}}(0)} \geq 0 \quad \text{and} \quad g_{|B_{\frac{4\gamma}{L_2}}(\frac{1}{2})} \leq 0\,. \tag{7.2}$$

Due to the Lipschitz-continuity of g by assumption (2.10), condition (7.2) can of course be ensured by choosing $\gamma \leq L_2/(4L_1)$.

Further, we remark that the symmetry condition (2.19) reduces the possible alternatives in Theorem 3.2 and leads to the following corollary:

COROLLARY 7.1 (Corollary 4.3 in [**39**]). *Suppose T satisfies all assertions of Theorem 3.2 and has the symmetry given by (2.19). Then one of the following holds:*

(i) *There exists one invariant graph φ with $\lambda(\varphi) \leq 0$. If φ has a negative Lyapunov exponent, it is always continuous. Otherwise the equivalence class contains at least an upper and a lower semi-continuous representative.*

(ii) *There exist three invariant graphs $\varphi^- \leq \psi \leq \varphi^+$ with $\lambda(\varphi^-) = \lambda(\varphi^+) < 0$ and $\lambda(\psi) > 0$. φ^- is always lower semi-continuous and φ^+ is always upper semi-continuous. Further, if one of the three graphs is continuous then so are the other two, if none of them is continuous there holds*

$$\overline{\Phi^-}^{ess} = \overline{\Psi}^{ess} = \overline{\Phi^+}^{ess}\,.$$

In addition, there holds

$$\varphi^-(\theta) = -\varphi^+(\theta + \tfrac{1}{2})$$

and

$$\psi(\theta) = -\psi(\theta + \tfrac{1}{2})\,.$$

Consequently, if we can show that there exists an SNA in a system of this kind, then we are automatically in situation (ii). Thus there will be two symmetric strange non-chaotic attractors which embrace a self-symmetric strange non-chaotic repellor, as claimed in Theorem 2.10 .

In order to repeat the construction from Section 6 for the case of symmetric forcing, we have to define admissible times and the sets R_N again. However, this time there are two critical intervals instead of one, namely $B_{\frac{4\gamma}{L_2}}(0)$ and $B_{\frac{4\gamma}{L_2}}(\frac{1}{2})$, corresponding to the maximum and minimum of the forcing function g. Therefore, we have to modify Definition 5.7 in the following way:

DEFINITION 7.2. *For $p \in \mathbb{N}_0 \cup \{\infty\}$ let $Q_p : \mathbb{Z} \to \mathbb{N}_0$ be defined by*

$$Q_p(j) := \begin{cases} q & \text{if } d(\omega_j, \{0, \frac{1}{2}\}) \in \left[S_{p-q+1}(\alpha) \cdot \frac{\alpha^{-q}}{L_2}, S_{p-q+2}(\alpha) \cdot \frac{\alpha^{-(q-1)}}{L_2}\right) \quad \text{for } q \geq 2 \\ 1 & \text{if } d(\omega_j, \{0, \frac{1}{2}\}) \in \left[S_p(\alpha) \cdot \frac{\alpha^{-1}}{L_2}, \frac{4\gamma}{L_2} + S_p(\alpha) \cdot \frac{\alpha^{-1}}{L_2} \cdot (1 - \mathbf{1}_{\{0\}}(p))\right) \\ 0 & \text{if } d(\omega_j, \{0, \frac{1}{2}\}) \geq \frac{4\gamma}{L_2} + S_p(\alpha) \cdot \frac{\alpha^{-1}}{L_2} \cdot (1 - \mathbf{1}_{\{0\}}(p)) \end{cases} .$$

if $j \in \mathbb{Z} \setminus \{0\}$ and $Q_p(0) := 0$. Again, let $p(j) := Q_0(j)$. Further let

$$\tilde{\nu}(q) := \min\left\{j \in \mathbb{N} \mid d(\omega_j, \{0, \tfrac{1}{2}\}) < 3S_\infty(\alpha) \cdot \frac{\alpha^{-(q-1)}}{L_2}\right\} \qquad \text{if } q \geq 2 \text{ and}$$

$$\tilde{\nu}(1) := \min\left\{j \in \mathbb{N} \mid d(\omega_j, \{0, \tfrac{1}{2}\}) < 3\left(\frac{4\gamma}{L_2} + S_\infty(\alpha) \cdot \frac{\alpha^{-1}}{L_2}\right)\right\} .$$

Apart from this, we define all the quantities $\Omega_p^{(\pm)}(j), \Omega^{(\pm)}, \tilde{\Omega}^{(\pm)}$ and ν exactly in the same way as in Definition 5.7, only using the altered definitions of the functions Q_p. Finally, we let

$$s(j) := \begin{cases} 1 & \text{if } d(\omega_j, 0) \leq \frac{4\gamma}{L_2} + \frac{S_\infty(\alpha)}{\alpha \cdot L_2} \\ -1 & \text{if } d(\omega_j, \frac{1}{2}) \leq \frac{4\gamma}{L_2} + \frac{S_\infty(\alpha)}{\alpha \cdot L_2} \\ 0 & \text{otherwise} \end{cases} .$$

In other words, we have just replaced $d(\omega_j, 0)$ by $d(\omega_j, \{0, \frac{1}{2}\})$ and introduced the function s in order to tell whether ω_j is close to 0 or to $\frac{1}{2}$. However, if we let $\tilde{\omega} := 2\omega \bmod 1$ there holds

$$d(\omega_j, \{0, \tfrac{1}{2}\}) = \tfrac{1}{2} d(\tilde{\omega}_j, 0) .$$

This means that Definition 5.7 with $\tilde{\omega}$ and $\tilde{L}_2 := \frac{1}{2}L_2$ yields exactly the same objects as Definition 7.2 with ω and L_2. Therefore, if we define all the quantities $l_q^{\pm}$, $J(m)$, A_N, Λ_N, R_N, Γ_N, ect. exactly in the same way as in Section 5, only with respect to Definition 7.2 instead of Definition 5.7, then all the results from Sections 5.2–5.4 will literally stay true. The only exception is Lemma 5.26(b), where we can even replace $d(\omega_j, 0)$ by $d(\omega_j, \{0, \frac{1}{2}\})$. Further, in Section 5.1 we did not use any specific assumption on g apart from the Lipschitz-continuity. Thus, we have all the tools from Section 5 available again.

Therefore, the only difference to the preceding section is the fact that the mapping $\beta \mapsto \xi_n(\beta, l)$ is not necessarily monotone anymore (where the $\xi_n(\beta, l)$ are defined exactly as before, see Definition 4.1). Hence, instead of considering arbitrary β as in Induction statement 6.1 we have to restrict to certain intervals $I_n^q = [\beta_{q,n}^+, \beta_{q,n}^-]$ ($q \in \mathbb{N}_0$, $n \in [l_q^+ + 1, \nu(q+1)]$ admissible) on which the dependence of $\xi_n(\beta, l_q^-)$ on β is monotone. The parameters $\beta_{q,n}^{\pm}$ will again satisfy

$$\xi_n(\beta_{q,n}^+, l_q^-) = \frac{1}{\alpha} \tag{7.3}$$

and

$$\xi_n(\beta^-_{q,n}, l^-_q) \;=\; -\frac{1}{\alpha}\;, \tag{7.4}$$

but they cannot be uniquely defined by these equations anymore.

The fact which makes up for the lack of monotonicity, and for the existence of the second critical region $B_{\frac{4\gamma}{L_2}}(\frac{1}{2})$, is that by deriving information about the orbits $\xi_n(\beta, l)$ we get another set of reference orbits for free: It follows directly from (2.19) that

$$\zeta_n(\beta, l) \;:=\; T_{\beta,-\omega_l+\frac{1}{2},n+l}(-3) \;=\; -\xi_n(\beta, l) \tag{7.5}$$

(Similar as in Definition 4.1, the $\zeta_n(\beta, l)$ correspond to the forward orbit of the points $(\omega_{-l} + \frac{1}{2}, -3)$, where we suppress the θ-coordinates again). Consequently, we have

$$\xi_n(\beta, l) \in \overline{B_{\frac{1}{\alpha}}(0)} \;\Leftrightarrow\; \zeta_n(\beta, l) \in \overline{B_{\frac{1}{\alpha}}(0)} \tag{7.6}$$

and

$$\xi_n(\beta, l) \geq \gamma \;\Leftrightarrow\; \zeta_n(\beta, l) \leq -\gamma\;. \tag{7.7}$$

In the case of symmetric forcing the induction statement reads as follows:

INDUCTION SCHEME 7.3. *Suppose the assumptions of Theorem 2.10 are satisfied and in addition (5.1), (5.1), (5.2), (6.1)–(6.4), (7.1) and (7.2) hold.*

Then for any $q \in \mathbb{N}_0$ and all admissible $n \in [l^+_q + 1, \nu(q+1)]$ there exists an interval $I^q_n = [\beta^+_{q,n}, \beta^-_{q,n}]$, such that $\beta^\pm_{q,n}$ satisfy (7.3) and (7.4) and in addition

I. $\beta \in I^q_{l^+_q+1}$ *implies*

$$\xi_j(\beta, l^-_q) \;\geq\; \gamma \quad \forall j \in [-l^-_q, 0] \setminus \Omega_\infty\;. \tag{7.8}$$

Further $I^q_{l^+_q+1} \subseteq \left[1 + \frac{1}{\sqrt{\alpha}}, 1 + \frac{3}{\sqrt{\alpha}}\right]$.

II. *For each admissible $n \in [l^+_q + 1, \nu(q+1)]$ the mapping $\beta \mapsto \xi_n(\beta, l^-_q)$ is strictly monotonically decreasing on I^q_n, (7.8) holds for all $\beta \in I^q_n$ and*

$$I^q_n \;\subseteq\; I^q_j \;\subseteq\; \left[1 + \frac{1}{\sqrt{\alpha}}, 1 + \frac{3}{\sqrt{\alpha}}\right] \qquad \forall j \in A_n \cap [l^+_q + 1, n]\;. \tag{7.9}$$

Further, for any $\beta \in I^q_n$ there holds

$$\xi_j(\beta, l^-_q) \;\in\; \overline{B_{\frac{1}{\alpha}}(0)} \qquad \forall j \in R_n\;. \tag{7.10}$$

III. *(a) If $n_1 \in [l^+_q + 1, \nu(q+1)]$ for some $q \geq 1$ there holds*

$$|\beta^+_{q,n_1} - \beta^-_{q,n_1}| \;\leq\; 2\alpha^{-\frac{n_1}{4}}\;. \tag{7.11}$$

In particular, in combination with (7.9) this implies that

$$|\beta^\pm_{q,n_1} - \beta| \;\leq\; 2\alpha^{-\frac{n_1}{4}} \qquad \forall \beta \in I^q_{n_2} \tag{7.12}$$

whenever $n_2 \in [l^+_q + 1, \nu(q+1)]$ and $n_1 \in A_{n_2}$ (as $I^q_{n_2} \subseteq I^q_{n_1}$ in this case).

(b) Let $1 \leq q_1 < q$, $n_1 \in [l^+_{q_1} + 1, \nu(q_1+1)]$ and $n_2 \in [l^+_q + 1, \nu(q) + 1]$. Then

$$|\beta^+_{q_1,n_1} - \beta^+_{q,n_2}| \;\leq\; 3 \cdot \sum_{i=q_1+1}^{q} \alpha^{-i} \;\leq\; \alpha^{-q_1}\;. \tag{7.13}$$

Theorem 2.10 now follows in exactly the same way as Theorem 2.7 from Induction scheme 6.1 (we do not repeat the argument here). The additional statements about the symmetry follow from Corollary 7.1 .

However, due to the lack of monotonicity we are not able to derive any further information about the sink-source-orbit or the bifurcation scenario as in the case of one-sided forcing. In particular, we have to leave open here whether β_0 is the only parameter at which an SNA occurs, or if this does indeed happen over a small parameter interval as the numerical observations suggest (compare Section 1.3.5).

7.1. Proof of the induction scheme

Standing assumption: In this whole subsection, we always assume that the assumptions of Induction scheme 7.3 are satisfied.

In order to start the induction we will need the following equivalent to Lemma 4.3, which can be proved in exactly in the same way (using that $d(\omega_j, 0) \geq \frac{3\gamma}{L_2}$ implies $g(\omega_j) \leq 1 - 3\gamma$ by (2.18) and (7.2), and similarly $d(\omega_j, \frac{1}{2}) \geq \frac{3\gamma}{L_2}$ implies $g(\omega_j) \geq -(1-3\gamma)$).

LEMMA 7.4. *Suppose that $\beta \leq 1 + \frac{4}{\sqrt{\alpha}}$ and $j \geq -l$. If $d(\omega_j, 0) \geq \frac{3\gamma}{L_2}$, then $\xi_j(\beta, l) \geq \frac{1}{\alpha}$ implies $\xi_{j+1}(\beta, l) \geq \gamma$. Similarly, if $d(\omega_j, \frac{1}{2}) \geq \frac{3\gamma}{L_2}$ then $\xi_j(\beta, l) \leq -\frac{1}{\alpha}$ implies $\xi_{j+1}(\beta, l) \leq -\gamma$. Consequently, $\xi_{j+1}(\beta, l) \in \overline{B_{\frac{1}{\alpha}}(0)}$ implies $\xi_j(\beta, l) \in \overline{B_{\frac{1}{\alpha}}(0)}$ whenever $d(\omega_j, \{0, \frac{1}{2}\}) \geq \frac{3\gamma}{L_2}$.*

Further, the following lemma replaces Lemma 6.2 . It will be needed to derive the required estimates on the parameters $\beta^{\pm}_{q,n}$ as well as the monotonicity of $\beta \mapsto \xi_n(\beta, l^-_q)$ on I^q_n.

LEMMA 7.5. *Let $q \in \mathbb{N}$ and let $n \in [l^+_q + 1, \nu(q+1)]$ be admissible. Further, assume*

$$\xi_j(\beta, l^-_q) \; \geq \; \gamma \qquad \forall j \in [-l^-_q, 0] \setminus \Omega_\infty \tag{7.1}$$

and

$$\xi_j(\beta, l^-_q) \; \in \; \overline{B_{\frac{1}{\alpha}}(0)} \qquad \forall j \in R_n \setminus \{n\} \; . \tag{7.2}$$

Then

$$\frac{\partial}{\partial \beta} \xi_n(\beta, l^-_q) \; \leq \; -\alpha^{\frac{n-1}{4}} \; . \tag{7.3}$$

PROOF. We have

$$\frac{\partial}{\partial \beta} \xi_{j+1}(\beta, l^-_q) \; = \; F'(\xi_j(\beta, l^-_q)) \cdot \frac{\partial}{\partial \beta} \xi_j(\beta, l^-_q) - g(\omega_j) \tag{7.4}$$

(compare (6.1)). In order to prove (7.3) we first have to obtain a suitable upper bound on $|\frac{\partial}{\partial \beta} \xi_0(\beta, l^-_q)|$. Let

$$\eta(j) \; := \; \#([-j, -1] \cap \Omega_\infty) \; .$$

We claim that under assumption (7.1) and for any $l \in [0, l^-_q]$ there holds

$$\left| \frac{\partial}{\partial \delta} \xi_0(\beta, l^-_q) \right| \; \leq \; \left| \frac{\partial}{\partial \delta} \xi_{-l}(\beta, l^-_q) \right| \cdot \alpha^{-\frac{1}{2}(l - 5\eta(l))} + \sum_{j=0}^{l-1} \alpha^{-\frac{1}{2}(j - 5\eta(j))} \; . \tag{7.5}$$

As $\eta(j) \leq \frac{j}{10}$ by (5.2) and $\frac{\partial}{\partial\delta}\xi_{-l_q^-}(\beta, l_q^-) = 0$ by definition, this implies

$$\left|\frac{\partial}{\partial\delta}\xi_0(\beta, l_q^-)\right| \leq S_\infty(\alpha^{\frac{1}{4}}) .$$

Using the fact that $\xi_0(\beta, l_q^-) \geq \gamma$ by assumption, this further yields

$$\frac{\partial}{\partial\delta}\xi_1(\beta, l_q^-) =$$

$$(7.6) \qquad = F'(\xi_0(\beta, l_q^-)) \cdot \frac{\partial}{\partial\delta}\xi_0(\beta, l_q^-) - 1 \overset{(2.7)}{\leq} -1 + \frac{S_\infty(\alpha^{\frac{1}{4}})}{2\sqrt{\alpha}} \overset{(6.4)}{\leq} -\frac{1}{2} .$$

We prove (7.5) by induction on l. For $l = 0$ the statement is obvious. In order to prove the induction step $l \to l+1$, first suppose that $-(l+1) \notin \Omega_\infty$, such that $\eta(l+1) = \eta(l)$ and $\xi_{-(l+1)}(\beta, l_q^-) \geq \gamma$. Then, using (7.4) we obtain

$$\begin{aligned}
\left|\frac{\partial}{\partial\delta}\xi_0(\beta, l_q^-)\right| &\leq \left|\frac{\partial}{\partial\delta}\xi_{-l}(\beta, l_q^-)\right| \cdot \alpha^{-\frac{1}{2}(l-5\eta(l))} + \sum_{j=0}^{l-1} \alpha^{-\frac{1}{2}(j-5\eta(j))} \\
&= \left|F'(\xi_{-(l+1)}(\beta, l_q^-)) \cdot \frac{\partial}{\partial\delta}\xi_{-(l+1)}(\beta, l_q^-) - g(\omega_{-(l+1)})\right| \cdot \alpha^{-\frac{1}{2}(l-5\eta(l))} \\
&\quad + \sum_{j=0}^{l-1} \alpha^{-\frac{1}{2}(j-5\eta(j))} \\
&\overset{(2.7)}{\leq} \left(\alpha^{-\frac{1}{2}} \cdot \left|\frac{\partial}{\partial\delta}\xi_{-(l+1)}(\beta, l_q^-)\right| + 1\right) \cdot \alpha^{-\frac{1}{2}(l-5\eta(l))} + \sum_{j=0}^{l-1} \alpha^{-\frac{1}{2}(j-5\eta(j))} \\
&= \left|\frac{\partial}{\partial\delta}\xi_{-(l+1)}(\beta, l_q^-)\right| \cdot \alpha^{-\frac{1}{2}(l+1-5\eta(l+1))} + \sum_{j=0}^{l} \alpha^{-\frac{1}{2}(j-5\eta(j))} .
\end{aligned}$$

The case $\eta(l+1) = \eta(l)+1$ is treated similarly, using (2.5) instead of (2.7) (compare with the proof of Lemma 5.2). This proves (7.5), such that (7.6) holds.

Now we can turn to prove (7.3). For any $k \in \mathbb{N}$ let

$$\tau(k) := \#([1, k-1] \setminus R_n) .$$

We will show the following statement by induction on k:

$$(7.7) \qquad \frac{\partial}{\partial\delta}\xi_k(\beta, l_q^-) \leq -\frac{1}{2} \cdot \left(\frac{3\sqrt{\alpha}}{2}\right)^{k-1-5\tau(k)} \qquad \forall k \in [1, n] .$$

As $\tau(n) \leq \frac{n-1}{10}$ by Lemma 5.27(a), this implies (7.3) whenever $n \geq l_q^+ + 1$. Note that $l_q^+ \geq 3$ by (6.6) and $\tau(n) = 0$ for all $n \leq 10$.

For $k = 1$ the statement is true by (7.6). Suppose that (7.7) holds for some $k \geq 1$ and first assume that $\tau(k+1) = \tau(k)$. Then

$$\begin{aligned}
\frac{\partial}{\partial\delta}\xi_{k+1}(\beta, l_q^-) &= F'(\xi_k(\beta, l_q^-)) \cdot \frac{\partial}{\partial\delta}\xi_k(\beta, l_q^-) - g(\omega_k) \\
&\stackrel{(2.6)}{\leq} -2\sqrt{\alpha} \cdot \frac{1}{2} \cdot \left(\frac{3\sqrt{\alpha}}{2}\right)^{k-1-5\tau(k)} + 1 \\
&\stackrel{(*)}{\leq} -(2\sqrt{\alpha} - 2) \cdot \frac{1}{2} \cdot \left(\frac{3\sqrt{\alpha}}{2}\right)^{k-1-5\tau(k)} \\
&\stackrel{(2.2)}{\leq} -\frac{1}{2} \cdot \left(\frac{3\sqrt{\alpha}}{2}\right)^{k-5\tau(k+1)},
\end{aligned}$$

$(*)$ where $\tau(k) \leq \frac{k-1}{10}$ by Lemma 5.27(a) ensures that $\left(\frac{3\sqrt{\alpha}}{2}\right)^{k-1-5\tau(k)}$ is always larger than 1. The case $\tau(k+1) = \tau(k) + 1$ is treated similar again, using (2.5) instead of (2.6) (compare with the proof of Lemma 5.4). Thus we have proved (7.7) and thereby the lemma.

□

As in Section 6, in order to prove Induction scheme 7.3 we proceed in six steps. The overall strategy needs some slight modifications in comparison to the case of one-sided forcing, but in many cases the proofs of the required estimates stay literally the same. In such situations we will not repeat all the details, but refer to the corresponding passages of the previous section instead.

Step 1: *Proof of the statement for $q = 0$*

Part I: Recall that $l_0^- = l_0^+ = 0$ and note that $\xi_0(\beta, 0) = 3 \geq \gamma$ by definition, such that (7.8) holds automatically. As $\frac{\partial}{\partial\beta}\xi_1(\beta, 0) = -1$, we can construct the interval I_1^0 by uniquely defining $\beta_{0,1}^{\pm}$ via (7.3) and (7.4). Further, we have $\xi_1(\beta, 0) = F(3) - \beta$. Using (2.4) and (2.7), it is easy to check that

$$F(3) \in [x_\alpha, x_\alpha + \tfrac{2-\frac{2}{\sqrt{\alpha}}}{2\sqrt{\alpha}}] \subseteq [1 + \tfrac{1}{\sqrt{\alpha}} + \tfrac{1}{\alpha}, 1 + \tfrac{3}{\sqrt{\alpha}} - \tfrac{1}{\alpha}] .$$

Therefore $I_0^1 = [\beta_{0,1}^+, \beta_{0,1}^-]$ must be contained in $[1 + \frac{1}{\sqrt{\alpha}}, 1 + \frac{3}{\sqrt{\alpha}}]$.

Parts II: We proceed by induction on n. For $n = 1$ the statement follows from the above. Suppose we have defined the intervals $I_n^0 \subseteq [1 + \frac{1}{\sqrt{\alpha}}, 1 + \frac{3}{\sqrt{\alpha}}]$ with the stated properties for all $n \leq N$, $N \in [1, \nu(1) - 1]$. As $p(N) = 0$, Lemma 7.4 yields that

$$\xi_{N+1}(\beta_{0,N}^+, 0) > \frac{1}{\alpha} \quad \text{and} \quad \xi_{N+1}(\beta_{0,N}^-, 0) < -\frac{1}{\alpha} .$$

This means that we can find $\beta_{0,N+1}^{\pm}$ in I_N^0 which satisfy (7.3) and (7.4). Consequently $I_{N+1}^0 := [\beta_{0,N+1}^+, \beta_{0,N+1}^-] \subseteq I_N^0$. It then follows from Part II of the induction statement for N, that $I_{N+1}^0 \subseteq I_j^0 \ \forall j \in [1, N] = R_N$, in particular $I_{N+1}^0 \subseteq I_1^0 \subseteq [1 + \frac{1}{\sqrt{\alpha}}, 1 + \frac{3}{\sqrt{\alpha}}]$ (note that $A_N = [1, N]$ as $N \leq \nu(1)$). This proves (7.8) and (7.9).

In order to see (7.10) suppose that $\beta \in I^0_{N+1}$. Then $\xi_N(\beta,0) \in \overline{B_{\frac{1}{\alpha}}(0)}$ by the definition of $I^0_{N+1} \subseteq I^0_N$ above, and therefore $\xi_j(\beta,0) \in \overline{B_{\frac{1}{\alpha}}(0)}$ $\forall j \in [1,N] = R_N$ follows from Part II of the induction statement for N. Finally, we can now use Lemma 7.5 to see that

$$\frac{\partial}{\partial\beta}\xi_{N+1}(\beta,0) \;\leq\; -\alpha^{\frac{N}{4}} \; . \tag{7.8}$$

This ensures the monotonicity of $\beta \mapsto \xi_{N+1}(\beta,0)$.

As Part III of the induction statement is void for $q = 0$, this completes Step I.

■

It remains to prove the induction step. Assume that the statement of Induction scheme 7.3 holds for all $q \leq p-1$. As in Section 6.1, the next two steps will prove Part I of the induction statement for p. Further, we can again assume in Step 2 and 3 that

$$p \;\geq\; 2 \; . \tag{7.9}$$

For the case $p = 1$ note that the analogue of Lemma 4.2 holds again in the case of symmetric forcing, with $d(\omega_j,0)$ being replaced by $d(\omega_j,\{0,\frac{1}{2}\})$, and this already shows Part I for $p = 1$.

Step 2: *If $|\beta - \beta^+_{p-1,\nu(p)}| \leq \alpha^{-p}$, then $\xi_j(\beta,l^-_p) \geq \gamma$ $\forall j \in [-l^-_p,0] \setminus \Omega_\infty$.*

Actually, this follows in exactly the same way as Step 2 in Section 6.1 . The crucial observation is the fact that Lemma 6.3 literally stays true in the situation of this section. As we will also need the statement for the reversed inequalities in the later steps, we restate it here:

LEMMA 7.6. *Let $q \geq 1$, $l^*, l \geq 0$, $\beta^* \in \left[1+\frac{1}{\sqrt{\alpha}}, 1+\frac{3}{\sqrt{\alpha}}\right]$ and $|\beta - \beta^*| \leq 2\alpha^{-q}$. Suppose that m is admissible, $p(m) \geq q$ and either $k = 0$ or $p(k) \geq q$. Further, suppose*

$$\xi_j(\beta^*,l^*) \;\in\; \overline{B_{\frac{1}{\alpha}}(0)} \quad \forall j \in R_m$$

and $\xi_{m+k-l^-_q}(\beta,l) \geq \gamma$. Then

$$\{j \in [m-l^-_q,m] \mid \xi_{j+k}(\beta,l) < \gamma\} \;\subseteq\; \tilde{\Omega}_{q-2} \; .$$

Similarly, if $\xi_{m+k-l^-_q}(\beta,l) \leq -\gamma$ then

$$\{j \in [m-l^-_q,m] \mid \xi_{j+k}(\beta,l) > -\gamma\} \;\subseteq\; \tilde{\Omega}_{q-2} \; .$$

The application of this lemma in order to show the statement of Step 2 is exactly the same as in Section 6.1 . The proof of the lemma is the same as for Lemma 6.3, apart from two slight modifications: First of all, Lemma 7.4 has to be used instead of Lemma 4.3 . Secondly, in order to show (6.3) two cases have to be distinguished. If $s(k) = 1$ nothing changes at all. For the second case $s(k) = -1$ it suffices just to replace the reference orbit

$$x^1_1,\ldots,x^1_n \;:=\; \xi_{j^-_i-1}(\beta^*,l^*),\ldots,\xi_{j^+_i}(\beta^*,l^*)$$

which is used for the application of Lemma 5.6(a) by

$$x_1^1, \dots, x_n^1 \;:=\; \zeta_{j_i^- - 1}(\beta^*, l^*), \dots, \zeta_{j_i^+}(\beta^*, l^*) \; .$$

Then the reference orbit starts on the fibre $\omega_{j_i^- - 1} + \frac{1}{2}$, and is therefore $\frac{\alpha^{-(q-1)}}{L_2}$-close to the first fibre $\omega_{j_i^- - 1 + k}$ of the second orbit

$$x_1^2, \dots, x_n^2 \;:=\; \xi_{j_i^- - 1 + k}(\beta, l), \dots, \xi_{j_i^+ + k}(\beta, l) \; ,$$

such that the error term is sufficiently small again. Due to (7.6) and (7.7), all further details then stay exactly the same as in the case $s(m) = 1$. The reader should be aware that even though the reference orbit changed, the set of times R_m at which it stays in the expanding region is the same as before. This is all which is needed in order to verify the assumptions of Lemma 5.6(a), which completes the proof of the lemma. Finally, the additional statement for the reversed inequalities can be shown similarly.

■

Step 3: *Construction of* $I^p_{l_p^+ + 1} \subseteq B_{\alpha^{-p}}(\beta^+_{p-1,\nu(p)})$.

Similar as in Step 3 of Section 6.1, we define $\beta^* := \beta^+_{p-1,\nu(p)}$, $\beta^+ := \beta^* - \alpha^{-p}$ and $\beta^- := \beta^* + \alpha^{-p}$. It then follows that

$$\xi_{l_p^+ + 1}(\beta^+, l_p^-) \;>\; \frac{1}{\alpha} \qquad \textit{and} \qquad \xi_{l_p^+ + 1}(\beta^-, l_p^-) \;<\; -\frac{1}{\alpha} \; . \tag{7.10}$$

The proof is exactly the same as for Claim 6.4, with reversed inequalities for the case of β^-. This means that we can define the parameters $\beta^\pm_{p,l_p^+ + 1}$ by

$$\beta^-_{p,l_p^+ + 1} \;:=\; \min\left\{ \beta \in B_{\alpha^{-p}}(\beta^*) \mid \xi_{l_p^+ + 1}(\beta, l_p^-) = -\frac{1}{\alpha} \right\} \tag{7.11}$$

and

$$\beta^+_{p,l_p^+ + 1} \;:=\; \max\left\{ \beta \in B_{\alpha^{-p}}(\beta^*) \mid \beta < \beta^-_{p,l_p^+ + 1},\ \xi_{l_p^+ + 1}(\beta, l_p^-) = \frac{1}{\alpha} \right\} \; . \tag{7.12}$$

Step 2 then implies that (7.8) is satisfied for

$$I^p_{l_p^+ + 1} \;:=\; \left[\beta^+_{p,l_p^+ + 1}, \beta^-_{p,l_p^+ + 1}\right]$$

and as $\beta^* \in \left[1 + \frac{1}{\sqrt{\alpha}}, 1 + \frac{3}{\sqrt{\alpha}}\right]$ there holds

$$I^p_{l_p^+ + 1} \;\subseteq\; \left[1 + \tfrac{1}{\sqrt{\alpha}} - \alpha^{-p}, 1 + \tfrac{3}{\sqrt{\alpha}} + \alpha^{-p}\right] \; . \tag{7.13}$$

$I^p_{l_p^+ + 1} \;\subseteq\; \left[1 + \frac{1}{\sqrt{\alpha}}, 1 + \frac{3}{\sqrt{\alpha}}\right]$ will be shown after Step 4. Apart from this the proof of Part I for $q = p$ is complete.

■

The next three steps will prove Part II and III of the induction statement for $q = p$, proceeding by induction on $n \in [l_p^+ + 1, \nu(p)]$. Again we start the induction with $n = l_p^+ + 1$.

Step 4: *Proof of Part II for* $q = p$ *and* $n = l_p^+ + 1$.

Let $\beta^* := \beta^+_{p-1,\nu(p)}$ again. We will prove the following claim:

CLAIM 7.7. *Suppose $\beta \in B_{\alpha^{-p}}(\beta^*)$ and $\xi_{l_p^++1}(\beta, l_p^-) \in \overline{B_{\frac{1}{\alpha}}(0)}$. Then*

$$\xi_j(\beta, l_p^-) \ \in \ \overline{B_{\frac{1}{\alpha}}(0)} \qquad \forall j \in R_{l_p^++1} \ . \tag{7.14}$$

This follows more or less in the same way as Step 4 in Section 6.1. Before we give the details, let us see how this implies the statement of Part II for $q = p$ and $n = l_p^+ + 1$:

In Step 3 we have constructed $I^p_{l_p^++1} \subseteq B_{\alpha^{-p}}(\beta^*)$. Suppose $\beta \in B_{\alpha^{-p}}(\beta^*)$ and $\xi_{l_p^++1}(\beta, l_p^-) \in \overline{B_{\frac{1}{\alpha}}(0)}$. Then Step 2 and the above claim ensure that the assumptions (7.1) and (7.2) of Lemma 7.5 are satisfied, and we obtain that $\xi_{l_p^++1}(\beta, l_p^-)$ is decreasing in β. In particular, this applies to β^+_{p,l_p^++1}. Consequently, if we increase β starting at β^+_{p,l_p^++1}, then $\xi_{l_p^++1}(\beta, l_p^-)$ will decrease until it leaves the interval $\overline{B_{\frac{1}{\alpha}}(0)}$. Due to the definition in (7.11) this is exactly the case when β^-_{p,l_p^++1} is reached. This yields the required monotonicity on $I^p_{l_p^++1}$, and (7.10) then follows from the claim. Note that (7.8) is already ensured by Step 2.

The proof of Claim 7.7 is completely analogous to that of Step 4 in Section 6.1: It will follow in the same way from the the two claims below, which correspond to Claims 6.5 and 6.6 .

CLAIM 7.8. *Suppose $\beta \in B_{\alpha^{-p}}(\beta^*)$ and $\xi_{l_p^++1}(\beta, l_p^-) \in \overline{B_{\frac{1}{\alpha}}(0)}$. If $\xi_k(\beta, l_p^-) \geq \frac{2}{\alpha}$ for some $k \in R_{l_p^++1}$ then $\xi_{l_p^++1}(\beta, l_p^-) \ > \ \frac{1}{\alpha}$. Similarly, if $\xi_k(\beta, l_p^-) \leq -\frac{2}{\alpha}$ then $\xi_{l_p^++1}(\beta, l_p^-) \ < \ -\frac{1}{\alpha}$.*

For $\xi_k(\beta, l_p^-) \geq \frac{2}{\alpha}$ this can be shown exactly as Claim 6.5. In the case $\xi_k(\beta, l_p^-) \leq -\frac{2}{\alpha}$ it suffices just to reverse all inequalities. The analogue to Claim 6.6 holds as well:

CLAIM 7.9. *Suppose $\beta \in B_{\alpha^{-p}}(\beta^*)$ and $\xi_{l_p^++1}(\beta, l_p^-) \in \overline{B_{\frac{1}{\alpha}}(0)}$. If $k \in R_{l_p^++1}$, $k+1 \in \Gamma_{l_p^++1}$ and $\xi_k(\beta, l_p^-) \geq \frac{1}{\alpha}$, then there exists some $\tilde{k} \in R_{l_p^++1}$ with $\xi_{\tilde{k}}(\beta, l_p^-) \geq \frac{2}{\alpha}$. Similarly, if $\xi_k(\beta, l_p^-) \leq -\frac{1}{\alpha}$ then there exists some $\tilde{k} \in R_{l_p^++1}$ with $\xi_{\tilde{k}}(\beta, l_p^-) \leq -\frac{2}{\alpha}$.*

PROOF. In order to prove this, we can proceed as in the proof of Claim 6.6: Suppose first that $\xi_k(\beta, l_p^-) \geq \frac{1}{\alpha}$ and define m, t and q' in exactly the same way. As these definitions only depend on the set $R_{l_p^++1}$, which is the same as before, there is no difference so far. Only instead of (6.17) we obtain

$$d(\omega_t, \{0, \tfrac{1}{2}\}) \ \leq \ \frac{1}{4} \cdot \frac{\alpha^{-(p(m)+1)}}{L_2} \tag{7.15}$$

Now we can apply Lemma 7.6, in the same way as Lemma 6.3 was applied in order to obtain (6.22), to conclude that

$$\{j \in [-l^-_{p(m)}, 0] \mid \xi_{j+m+t}(\beta, l_p^-) < \gamma\} \ \subseteq \ \Omega_\infty \ . \tag{7.16}$$

For the further argument we have to distinguish two cases. If $s(m+t) = 1$, then we can use exactly the same comparison arguments as in Section 6.1 to show that

$\xi_{m+t+l^+_{p(m)}+1}(\beta, l^-_p) \geq \frac{2}{\alpha}$ if $p(m) \geq 2$. The details all remain exactly the same. Thus, we can choose $\tilde{k} = m+t+l^+_{p(m)}+1$ if $p(m) \geq 2$ and again $\tilde{k} = m+t+1$ or $m+t+2$ if $p(m) = 1$.

On the other hand, suppose $s(m+t) = -1$. Then $d(\omega_{m+t}, 0) \geq \frac{3\gamma}{L_2}$, and in addition (7.16) implies that $\xi_{m+t}(\beta, l^-_p) \geq \gamma$. Lemma 7.4 therefore yields that $\xi_{m+t+1}(\beta, l^-_p) \geq \gamma \geq \frac{2}{\alpha}$, such that we can choose $\tilde{k} = m+t+1$.

The case $\xi_k(\beta, l^-_p) \leq -\frac{1}{\alpha}$ is then treated analogously: First of all, application of Lemma 7.6 yields

$$\{j \in [-l^-_{p(m)}, 0] \mid \xi_{j+m+t}(\beta, l^-_p) > -\gamma\} \subseteq \Omega_\infty , \tag{7.17}$$

in particular $\xi_{m+t}(\beta, l^-_p) \leq -\gamma$. If $s(m+t) = 1$, such that $d(\omega_{m+t}, \frac{1}{2}) \geq \frac{3\gamma}{L_2}$, then Lemma 7.4 yields that $\xi_{m+t+1}(\beta, l^-_p) \leq -\gamma \leq -\frac{2}{\alpha}$ and we can choose $\tilde{k} = m+t+1$.

On the other hand, if $s(m+t) = -1$, then we can again apply similar comparison arguments as in the proof of Claim 6.6 to conclude that $\xi_{m+t+l^+_{p(m)}+1}(\beta, l^-_p) \leq -\frac{2}{\alpha}$ if $p(m) \geq 2$ (and $\xi_{m+t+1}(\beta, l^-_p) \leq -\frac{2}{\alpha}$ or $\xi_{m+t+2}(\beta, l^-_p) \leq -\frac{2}{\alpha}$ if $p(m) = 1$). Apart from the reversed inequalities, the only difference now is that the reference orbits $\xi_{-l^-_{p(m)}}(\beta^*, l^-_{q'}), \ldots, \xi_{-1}(\beta^*, l^-_{q'})$ and $\xi_1(\beta^*, l^-_{q'}), \ldots, \xi_{l^+_{p(m)}}(\beta^*, l^-_{q'})$ in (6.23) and (6.27) have to be replaced by $\zeta_{-l^-_{p(m)}}(\beta^*, l^-_{q'}), \ldots, \zeta_{-1}(\beta^*, l^-_{q'})$ and $\zeta_1(\beta^*, l^-_{q'}), \ldots, \zeta_{l^+_{p(m)}}(\beta^*, l^-_{q'})$, respectively. Due to (7.6) and (7.7), all other details remain exactly the same as before, with (2.11) being replaced by (2.18).

□

■

Now we can also show that $I^p_{l^+_p+1} \subseteq \left[1+\frac{1}{\sqrt{\alpha}}, 1+\frac{3}{\sqrt{\alpha}}\right]$, which completes the proof of Part I of the induction statement for p. Suppose that $\beta \in I^p_{l^+_p+1}$. Then, due to Step 2 and the construction of $I^p_{l^+_p+1} \subseteq B_{\alpha^{-p}}(\beta^+_{p-1,\nu(p)})$ in Step 3, (7.8) holds, such that in particular $\xi_0(\beta, l^-_p) \geq \gamma$. Thus, it follows from (2.4) and (2.7) that

$$\xi_1(\beta, l^-_p) \in \left[1+\frac{3}{2\sqrt{\alpha}} - \beta, 1+\frac{3}{\sqrt{\alpha}} - \frac{1}{\alpha} - \beta\right].$$

As Step 4 yields that $\xi_1(\beta, l^-_p) \in \overline{B_{\frac{1}{\alpha}}(0)}$ and $\frac{1}{2\sqrt{\alpha}} \geq \frac{1}{\alpha}$, this implies $\beta \in \left[1+\frac{1}{\sqrt{\alpha}}, 1+\frac{3}{\sqrt{\alpha}}\right]$ as required.

Step 5: *Part II of the induction statement implies Part III*

As in Section 6.1, we suppose that Part II with $q = p$ holds for all $n \leq N$, with $N \in [l^+_p+1, \nu(p+1)]$, and show that in this case Part III(a) holds as well whenever $n_1, n_2 \leq N$ and similarly Part III(b) holds whenever $n_2 \leq N$.

Let $n_1 \leq N$ be admissible. As we assume that Part II of the induction statement with $q = p$ holds for $n = n_1$, we can use Lemma 7.5 to see that $\frac{\partial}{\partial\beta}\xi_{n_1}(\beta, l^-_p) \leq$

$-\alpha^{\frac{n_1}{4}}$ for all $\beta \in I^p_{n_1}$, which implies (7.11). Then (7.12) is a direct consequence of (7.9). This proves Part III(a). Part III(b) follows in the same way as in Step 3 of Section 6.1 .

■

Step 6: *Proof of Part II for $q = p$.*

In order to prove Part II of the induction statement for $q = p$, we proceed by induction on n. In Step 4 we already constructed $I^p_{l_p^+ + 1}$ with the required properties. Now suppose that I^p_n has been constructed for all admissible $n \in [l_p^+ + 1, N]$, where $N \in [l_p^+ + 1, \nu(p+1) - 1]$. We now have to construct I^p_{N+1} with the required properties, provided $N+1$ is admissible. Again, the case where N is admissible as well is rather easy: In this case $p(N) = 0$, otherwise $N+1$ would be contained in $J(N)$. Therefore Lemma 7.4 yields that

$$\xi_{N+1}(\beta^+_{p,N}, l_p^-) > \frac{1}{\alpha} \tag{7.18}$$

and

$$\xi_{N+1}(\beta^-_{p,N}, l_p^-) < -\frac{1}{\alpha}. \tag{7.19}$$

Consequently, we can find $\beta^{\pm}_{p,N+1} \in I^p_N$ which satisfy (7.3) and (7.4), such that $I^p_{N+1} = [\beta^+_{p,N+1}, \beta^-_{p,N+1}] \subseteq I^p_N$. Note that $R_N = R_{N+1} \setminus \{N+1\}$ by (5.4). Therefore Part II of the induction statement for $n = N$ implies that we can apply Lemma 7.5 to any $\beta \in I^p_{N+1}$, and this yields the monotonicity of $\xi_{N+1}(\beta, l_p^-)$ on I^p_{N+1}. All other required statements for $n = N+1$ then follow directly from Part II of the induction statement for $n = N$.

It remains to treat the case where $N+1$ is admissible but N is not admissible. As in Step 6 of Section 6.1 we have to consider the interval $J \in \mathcal{J}_{N+1}$ which contains N, i.e. $J = [t, N]$ with $t := \lambda^-(m_J)$. In order to construct I^p_{N+1} inside of I^p_{t-1} we prove the following claim (compare Claim 6.7):

CLAIM 7.10. *$\xi_{N+1}(\beta^+_{p,t-1}, l_p^-) > \frac{1}{\alpha}$ and $\xi_{N+1}(\beta^-_{p,t-1}, l_p^-) < -\frac{1}{\alpha}$.*

PROOF. We only give an outline here, the details can be checked exactly as in the proof of Claim 6.7 . Note that it sufficed there to show (6.30), such that the problem is analogous.

Let $\beta^+ := \beta^+_{p,t-1}$ and $m := m_J$. First, we can apply Lemma 7.6 with $q = p(m)$, $l = l^* = l_p^-$, $\beta^* = \beta^+_{p,m}$, m as above, $k = 0$ and $\beta = \beta^+$ to obtain that

$$\{j \in [-l^-_{p(m)}, 0] \mid \xi_{j+m}(\beta^+, l_p^-) < \gamma\} \subseteq \Omega_\infty \tag{7.20}$$

(compare (6.31)–(6.36)). Then we have to distinguish two cases. If $s(m) = 1$, we can proceed as in the proof of 6.7 to show that $\xi_{N+1}(\beta^+, l_p^-) \geq \frac{2}{\alpha}$. On the other hand suppose $s(m) = -1$, such that $d(\omega_{m_J}, 0) \geq \frac{4\gamma}{L_2}$. In this case (7.20) implies in particular that $\xi_m(\beta^+, l_p^-) \geq \gamma$, and Lemma 7.4 therefore yields $\xi_{m+1}(\beta^+, l_p^-) \geq \gamma \geq \frac{2}{\alpha}$. Similar to the case $s(m) = 1$ we can now compare the orbits

$$x^1_1, \ldots, x^1_n := \zeta_1(\beta^+, l_p^-), \ldots, \zeta_{l^+_{p(m)}}(\beta^+, l_p^-) \tag{7.21}$$

and

$$x^2_1, \ldots, x^2_n := \xi_{m+1}(\beta^+, l_p^-), \ldots, \xi_N(\beta^+, l_p^-), \tag{7.22}$$

(see (6.40) and (6.41)), with the difference that it suffices to use Lemma 5.6(a) instead of (b). Note that the information we have about the orbit (7.21) is exactly the same as for the orbit (6.40) (see (7.6)). Thus, we also obtain $\xi_{N+1}(\beta^+, l_p^-) > \frac{1}{\alpha}$ in this case.

The proof for $\xi_{N+1}(\beta^-, l_p^-) < -\frac{1}{\alpha}$ is then analogous. This time, it suffices to use Lemma 5.6(a) for the case $s(m) = 1$, whereas Lemma 5.6(b) has to be invoked in order to compare the orbits $x_1^1, \ldots, x_n^1 := \zeta_1(\beta^+, l_p^-), \ldots, \zeta_{l^+_{p(m)}}(\beta^+, l_p^-)$ and $x_1^2, \ldots, x_n^2 := \xi_{m+1}(\beta^+, l_p^-), \ldots, \xi_N(\beta^+, l_p^-)$ in case $s(m) = -1$, but the details for the application are again the same as before.

□

Using the above claim, we see that

$$\beta^-_{p,N+1} := \min\left\{\beta \in I^p_{t-1} \mid \xi_{N+1}(\beta, l_p^-) = -\frac{1}{\alpha}\right\} \tag{7.23}$$

and

$$\beta^+_{p,N+1} := \max\left\{\beta \in I^p_{t-1} \mid \beta < \beta^-_{p,N+1},\ \xi_{N+1}(\beta, l_p^-) = \frac{1}{\alpha}\right\} \tag{7.24}$$

are well defined, such that $I^p_{N+1} := [\beta^+_{p,N+1}, \beta^-_{p,N+1}] \subseteq I^p_{t-1}$. Then, due to Part II of the induction statement for $n = t-1$, (7.8) holds for all $\beta \in I^p_{N+1}$ and similarly

$$\xi_j(\beta, l_p^-) \in \overline{B_{\frac{1}{\alpha}}(0)} \quad \forall j \in R_{t-1} \tag{7.25}$$

whenever $\beta \in I^p_{N+1}$. As $R_{N+1} = R_{t-1} \cup R(J) \cup \{N+1\}$, it remains to obtain information about $R(J)$. Thus, in order to complete this step we need the following claim, which is the analog of Claim 6.8:

CLAIM 7.11. *Suppose $\beta \in I^p_{N+1}$ and $\xi_{N+1}(\beta, l_p^-) \in \overline{B_{\frac{1}{\alpha}}(0)}$. Then $\xi_j(\beta, l_p^-) \in \overline{B_{\frac{1}{\alpha}}(0)}$ $\forall j \in R(J)$.*

Similar to Claim 6.8, this follows from two further claims, which are the analogues of Claims 6.9 and 6.10 . Before we state them, let us see how we can use Claim 7.11 in order to complete the induction step $N \to N+1$ and thereby the proof of Step 6:

Suppose that $\beta \in I^p_{N+1}$ and $\xi_{N+1}(\beta, l_p^-) \in \overline{B_{\frac{1}{\alpha}}(0)}$. Then (7.25) together with the claim imply that

$$\xi_j(\beta, l_p^-) \in \overline{B_{\frac{1}{\alpha}}(0)} \quad \forall j \in R_{N+1} . \tag{7.26}$$

In addition (7.8) holds, as mentioned before (7.25). Consequently, Lemma 7.5 (with $q = p$ and $n = N+1$) implies that

$$\frac{\partial}{\partial \beta}\xi_{N+1}(\beta, l_p^-) \leq -\alpha^{\frac{N}{4}} .$$

In particular, this is true for $\beta = \beta^+_{p,N+1}$, and when β is increased it will remain true until $\xi_{N+1}(\beta, l_p^-)$ leaves $\overline{B_{\frac{1}{\alpha}}(0)}$, i.e. all up to $\beta^-_{p,N+1}$. This proves the required monotonicity of $\beta \mapsto \xi_{N+1}(\beta, l_p^-)$ on I^p_{N+1}, and thus Part II of the induction statement holds for $n = N+1$.

CLAIM 7.12. *Suppose $\xi_k(\beta, l_p^-) \geq \frac{2}{\alpha}$ for some $k \in R(J)$. Then $\xi_{N+1}(\beta, l_p^-) > \frac{1}{\alpha}$. Similarly, if $\xi_k(\beta, l_p^-) \leq -\frac{2}{\alpha}$ then $\xi_{N+1}(\beta, l_p^-) < -\frac{1}{\alpha}$.*

This is proved exactly as Claim 6.9, with all inequalities reversed for the case $\xi_k(\beta, l_p^-) \leq -\frac{2}{\alpha}$.

CLAIM 7.13. *Suppose $k \in R(J)$, $k+1 \in \Gamma^+(J)$ and $\xi_k(\beta, l_p^-) \geq \frac{1}{\alpha}$. Then there exists some $\tilde{k} \in R(J)$ with $\xi_{\tilde{k}}(\beta, l_p^-) \geq \frac{2}{\alpha}$. Similarly, if $\xi_k(\beta, l_p^-) \leq -\frac{1}{\alpha}$ there exists some $\tilde{k} \in R(J)$ with $\xi_{\tilde{k}}(\beta, l_p^-) \leq -\frac{2}{\alpha}$.*

PROOF. This can be shown in the same way as Claim 6.10: Suppose first that $\xi_k(\beta, l_p^-) \geq \frac{1}{\alpha}$ and define m, t and q' as in the proof of Claim 6.10. As these definitions only depend on the sets of regular points, which are the same as before, there is no difference so far. Only instead of (6.46) we obtain

$$d(\omega_t, \{0, \tfrac{1}{2}\}) \leq \frac{1}{4} \cdot \frac{\alpha^{-(p(m)+1)}}{L_2} \tag{7.27}$$

Nevertheless, we can apply Lemma 7.6, in the same way as Lemma 6.3 was applied in order to obtain (6.51), to conclude that

$$\{j \in [-l_{p(m)}^-, 0] \mid \xi_{j+m+t}(\beta, l_p^-) < \gamma\} \subseteq \Omega_\infty \tag{7.28}$$

(compare (6.47)–(6.51)). For the further argument we have to distinguish two cases. If $s(m+t) = 1$ and $p(m) \geq 2$, then we can use exactly the same comparison arguments as for Claim 6.10 (compare (6.52)–(6.58)) to show that $\xi_{m+t+l_{p(m)}^+ +1}(\beta, l_p^-) \geq \frac{2}{\alpha}$. The details all remain exactly the same. Thus, we can choose $\tilde{k} = m+t+l_{p(m)}^+ +1$ if $p(m) \geq 2$, and similarly $\tilde{k} = m+t+1$ or $m+t+2$ if $p(m) = 1$.

On the other hand, suppose $s(m+t) = -1$. Then $d(\omega_{m+t}, 0) \geq \frac{3\gamma}{L_2}$, and in addition (7.28) implies that $\xi_{m+t}(\beta, l_p^-) \geq \gamma$. Lemma 7.4 therefore yields that $\xi_{m+t+1}(\beta, l_p^-) \geq \gamma \geq \frac{2}{\alpha}$, such that we can choose $\tilde{k} = m+t+1$.

The case $\xi_k(\beta, l_p^-) \leq -\frac{1}{\alpha}$ is then treated analogously: First of all, application of Lemma 7.6 yields

$$\{j \in [-l_{p(m)}^-, 0] \mid \xi_{j+m+t}(\beta, l_p^-) > -\gamma\} \subseteq \Omega_\infty , \tag{7.29}$$

in particular $\xi_{m+t}(\beta, l_p^-) \leq -\gamma$. If $s(m+t) = 1$, such that $d(\omega_{m+t}, \frac{1}{2}) \geq \frac{3\gamma}{L_2}$, then Lemma 7.4 yields that $\xi_{m+t+1}(\beta, l_p^-) \leq -\gamma \leq -\frac{2}{\alpha}$ and we can choose $\tilde{k} = m+t+1$.

On the other hand, if $s(m+t) = -1$, then we can again apply similar comparison arguments as in the proof of Claim 6.10 (compare (6.52)–(6.58))to conclude that $\xi_{m+t+l_{p(m)}^+ +1}(\beta, l_p^-) \leq -\frac{2}{\alpha}$ if $p(m) \geq 2$ (and again $\xi_{m+t+1}(\beta, l_p^-) \leq -\frac{2}{\alpha}$ or $\xi_{m+t+2}(\beta, l_p^-) \leq -\frac{2}{\alpha}$ if $p(m) = 1$). Apart from the reversed inequalities the only difference now is that the reference orbits $\xi_{-l_{p(m)}^-}(\beta^*, l_{q'}^-), \ldots, \xi_{-1}(\beta^*, l_{q'}^-)$ and $\xi_1(\beta^*, l_{q'}^-), \ldots, \xi_{l_{p(m)}^+}(\beta^*, l_{q'}^-)$ in (6.52) and (6.56) have to be replaced by $\zeta_{-l_{p(m)}^-}(\beta^*, l_{q'}^-)$, $\ldots$, $\zeta_{-1}(\beta^*, l_{q'}^-)$ and $\zeta_1(\beta^*, l_{q'}^-)$, $\ldots$, $\zeta_{l_{p(m)}^+}(\beta^*, l_{q'}^-)$, respectively. Due to (7.6) and (7.7), all other details remain exactly the same as before.

□

■

Bibliography

[1] Michael R. Herman. Une méthode pour minorer les exposants de Lyapunov et quelques exemples montrant le caractère local d'un théorème d'Arnold et de Moser sur le tore de dimension 2. *Commentarii Mathematici Helvetici*, 58:453–502, 1983.

[2] Celso Grebogi, Edward Ott, Steven Pelikan, and James A. Yorke. Strange attractors that are not chaotic. *Physica D*, 13:261–268, 1984.

[3] John Milnor. On the concept of attractor. *Communications in Mathematical Physics*, 99:177–195, 1985.

[4] Gerhard Keller. A note on strange nonchaotic attractors. *Fundamenta Mathematicae*, 151(2):139–148, 1996.

[5] Hinke Osinga, Jan Wiersig, Paul Glendinning, and Ulrike Feudel. Multistability and non-smooth bifurcations in the quasiperiodically forced circle map. *International Journal of Bifurcation and Chaos*, 11(12):3085–3105, 2001.

[6] Anders Bondeson, Edward Ott, and Thomas M. Antonsen, Jr. Quasiperiodically forced damped pendula and Schrödinger equations with quasiperiodic potentials: Implications of their equivalence. *Physical Review Letters*, 55(20):2103–2106, 1985.

[7] Filipe J. Romeiras and Edward Ott. Strange nonchaotic attractors of the damped pendulum with quasiperiodic forcing. *Physical Review A*, 35(10):4404–4413, 1987.

[8] Filipe J. Romeiras, Anders Bodeson, Edward Ott, Thomas M. Antonsen Jr., and Celso Grebogi. Quasiperiodically forced dynamical systems with strange nonchaotic attractors. *Physica D*, 26:277–294, 1987.

[9] Mingzhou Ding, Celso Grebogi, and Edward Ott. Evolution of attractors in quasiperiodically forced systems: From quasiperiodic to strange nonchaotic to chaotic. *Physical Review A*, 39(5):2593–2598, 1989.

[10] J.F. Heagy and S.M. Hammel. The birth of strange nonchaotic attractors. *Physica D*, 70:140–153, 1994.

[11] Arkady S. Pikovski and Ulrike Feudel. Characterizing strange nonchaotic attractors. *Chaos*, 5(1):253–260, 1995.

[12] Ulrike Feudel, Jürgen Kurths, and Arkady S. Pikovsky. Strange nonchaotic attractor in a quasiperiodically forced circle map. *Physica D*, 88:176–186, 1995.

[13] Annette Witt, Ulrike Feudel, and Arkady Pikovsky. Birth of strange nonchaotic attractors due to interior crisis. *Physica D*, 109:180–190, 1997.

[14] P.R. Chastell, P.A. Glendinning, and J. Stark. Locating bifurcations in quasiperiodically forced systems. *Physics Letters A*, 200:17–26, 1995.

[15] Paul Glendinning. Intermittency and strange nonchaotic attractors in quasiperiodically forced circle maps. *Physics Letters A*, 244:545–550, 1998.

[16] Rob Sturman. Scaling of intermittent behaviour of a strange nonchaotic attractor. *Physics Letters A*, 259(5):355–365, 1999.

[17] P. Glendinning, U. Feudel, A.S. Pikovsky, and J. Stark. The structure of mode-locked regions in quasi-periodically forced circle maps. *Physica D*, 140:227–243, 2000.

[18] Surendra Singh Negi, Awadhesh Prasad, and Ramakrishna Ramaswamy. Bifurcations and transitions in the quasiperiodically forced logistic map. *Physica D*, 145:1–12, 2000.

[19] Awadhesh Prasad, Surendra Singh Negi, and Ramakrishna Ramaswamy. Strange nonchaotic attractors. *International Journal of Bifurcation and Chaos*, 11(2):291–309, 2001.

[20] Haro, A. and de la Llave, R. Manifolds at the verge of a hyperbolicity breakdown. *Chaos* 16(1):21–30, 2006.

[21] W. L. Ditto, M. L. Spano, H. T. Savage, S. N. Heagy J. Rauseo, and E. Ott. Experimental observation of a strange nonchaotic attractor. *Physical Review Letters*, 65(5):533–536, 1990.

[22] J. Heagy and W. L. Ditto. Dynamics of a two-frequency parametrically driven Duffing oscillator. *Journal of Nonlinear Science*, 1(4):423–455, 1991.

[23] Ting Zhou, Frank Moss, and Adi Bulsara. Observation of a strange nonchaotic attractor in a multistable potential. *Physical Review A*, 45(8):5394–5401, 1992.

[24] J. Stark. Regularity of invariant graphs for forced systems. *Ergodic Theory and Dynamical Systems*, 19(1):155–199, 1999.

[25] J. Stark and R. Sturman. Semi-uniform ergodic theorems and applications to forced systems. *Nonlinearity*, 13(1):113–143, 2000.

[26] T. Jäger and G. Keller. The denjoy type-of argument for quasiperiodically forced circle diffeomorphisms. *To appear in* Ergodic Theory and Dynamical Systems, 2004.

[27] R. Fabbri, T. Jäger, R. Johnson, and G. Keller. A Sharkovskii-type theorem for minimally forced interval maps. *Topological Methods in Nonlinear Analysis*, 26:163–188, 2005.

[28] T. Jäger and J. Stark. Towards a classification of quasiperiodically forced circle homeomorphisms. *To appear in* Journal of the LMS, 2005.

[29] A. Avila and R. Krikorian. Reducibility or non-uniform hyperbolicity for quasiperiodic Schrödinger cocycles. To appear in Annals of Mathematics.

[30] A. Avila and Jitomirskaya. The ten martini problem. Preprint.

[31] Kristian Bjerklöv. Positive lyapunov exponent and minimality for a class of one-dimensional quasi-periodic schrödinger equations. *Ergodic Theory and Dynamical Systems*, 25:1015–1045, 2005.

[32] Kristian Bjerklöv. *Dynamical Properties of Quasi-periodic Schrödinger equations.* PhD thesis, Royal Institute of Technology, Stockholm, 2003.

[33] Kristian Bjerklöv. Dynamics of the quasiperiodic Schrödinger cocycle at the lowest energy in the spectrum. Preprint 2005.

[34] Paul Glendinning. Global attractors of pinched skew products. *Dynamical Systems*, 17:287–294, 2002.

[35] T. Jäger. On the structure of strange nonchaotic attractors in pinched skew products. *To appear in* Ergodic theory and Dynamical systems, 2004.

[36] S. Datta, T. Jäger, G. Keller, and R. Ramaswamy. On the dynamics of the critical harper map. *Nonlinearity*, 17:2315–2323, 2004.

[37] J. Stark. Transitive sets for quasiperiodically forced monotone maps. *Dynamical Systems*, 18(4):351–364, 2003.

[38] L. Arnold. *Random Dynamical Systems.* Springer, 1998.

[39] T. Jäger. Quasiperiodically forced interval maps with negative Schwarzian derivative. *Nonlinearity*, 16(4):1239–1255, 2003.

[40] Rufus Bowen. Entropy for group endomorphisms and homogeneous spaces. *Transactions of the AMS*, 153:401–413, 1971.

[41] A. Katok and B. Hasselblatt. *Introduction to the Modern Theory of Dynamical Systems.* Cambridge University Press, 1995.

[42] W. de Melo and S. van Strien. *One-dimensional dynamics.* Springer, 1993.

[43] A. Haro and J. Puig. Strange non-chaotic attractors in Harper maps. *Chaos* 16, 033127, 2006.

[44] S. Y. Jitomirskaya. Metal-insulator transition for the almost Mathieu operator. *Annals of Mathematics (2)*, 150(3):1159–1175, 1999.

[45] M. Benedicks and L. Carleson. The dynamics of the Hénon map. *Annals of Mathematics (2)*, 133(1):73–169, 1991.

[46] Puig, J. A nonperturbative Eliasson's reducibility theorem. *Nonlinearity*, 19:355–376, 2006.

[47] Aubry, S. and André, G. Analyticity breaking and Anderson localization in incommensurate lattices. *Proc. 8th Int. Coll. on Group Theoretical Methods in Physics, Kiryat Anavim, Israel 1979 (Bristol: Hilger).* Ann. Isr. Phys. Soc. 3:133–164, 1980.

[48] P. Glendinning. Non-smooth pitchfork bifurcations. *Discrete and Continuous Dynamical Systems*, B 4(2):457–464, 2004.

Editorial Information

To be published in the *Memoirs*, a paper must be correct, new, nontrivial, and significant. Further, it must be well written and of interest to a substantial number of mathematicians. Piecemeal results, such as an inconclusive step toward an unproved major theorem or a minor variation on a known result, are in general not acceptable for publication.

Papers appearing in *Memoirs* are generally at least 80 and not more than 200 published pages in length. Papers less than 80 or more than 200 published pages require the approval of the Managing Editor of the Transactions/Memoirs Editorial Board.

As of May 31, 2009, the backlog for this journal was approximately 11 volumes. This estimate is the result of dividing the number of manuscripts for this journal in the Providence office that have not yet gone to the printer on the above date by the average number of monographs per volume over the previous twelve months, reduced by the number of volumes published in four months (the time necessary for preparing a volume for the printer). (There are 6 volumes per year, each usually containing at least 4 numbers.)

A Consent to Publish and Copyright Agreement is required before a paper will be published in the *Memoirs*. After a paper is accepted for publication, the Providence office will send a Consent to Publish and Copyright Agreement to all authors of the paper. By submitting a paper to the *Memoirs*, authors certify that the results have not been submitted to nor are they under consideration for publication by another journal, conference proceedings, or similar publication.

Information for Authors

Memoirs are printed from camera copy fully prepared by the author. This means that the finished book will look exactly like the copy submitted.

Initial submission. The AMS uses Centralized Manuscript Processing for initial submissions. Authors should submit a PDF file using the Initial Manuscript Submission form found at `www.ams.org/peer-review-submission`, or send one copy of the manuscript to the following address: Centralized Manuscript Processing, MEMOIRS OF THE AMS, 201 Charles Street, Providence, RI 02904-2294 USA. If a paper copy is being forwarded to the AMS, indicate that it is for it Memoirs and include the name of the corresponding author, contact information such as email address or mailing address, and the name of an appropriate Editor to review the paper (see the list of Editors below).

The paper must contain a *descriptive title* and an *abstract* that summarizes the article in language suitable for workers in the general field (algebra, analysis, etc.). The *descriptive title* should be short, but informative; useless or vague phrases such as "some remarks about" or "concerning" should be avoided. The *abstract* should be at least one complete sentence, and at most 300 words. Included with the footnotes to the paper should be the 2000 *Mathematics Subject Classification* representing the primary and secondary subjects of the article. The classifications are accessible from `www.ams.org/msc/`. The list of classifications is also available in print starting with the 1999 annual index of *Mathematical Reviews*. The Mathematics Subject Classification footnote may be followed by a list of *key words and phrases* describing the subject matter of the article and taken from it. Journal abbreviations used in bibliographies are listed in the latest *Mathematical Reviews* annual index. The series abbreviations are also accessible from `www.ams.org/msnhtml/serials.pdf`. To help in preparing and verifying references, the AMS offers MR Lookup, a Reference Tool for Linking, at `www.ams.org/mrlookup/`.

Electronically prepared manuscripts. The AMS encourages electronically prepared manuscripts, with a strong preference for $\mathcal{A}\mathcal{M}\mathcal{S}$-LaTeX. To this end, the Society has prepared $\mathcal{A}\mathcal{M}\mathcal{S}$-LaTeX author packages for each AMS publication. Author packages include instructions for preparing electronic manuscripts, samples, and a style file that generates the particular design specifications of that publication series. Though $\mathcal{A}\mathcal{M}\mathcal{S}$-LaTeX is the highly preferred format of TeX, author packages are also available in $\mathcal{A}\mathcal{M}\mathcal{S}$-TeX.

Authors may retrieve an author package for *Memoirs of the AMS* from `www.ams.org/journals/memo/memoauthorpac.html` or via FTP to `ftp.ams.org` (login as `anonymous`, enter username as password, and type `cd pub/author-info`). The *AMS Author Handbook* and the *Instruction Manual* are available in PDF format from the author package link. The author package can also be obtained free of charge by sending email to `tech-support@ams.org` (Internet) or from the Publication Division, American Mathematical Society, 201 Charles St., Providence, RI 02904-2294, USA. When requesting an author package, please specify $\mathcal{A}\mathcal{M}\mathcal{S}$-LaTeX or $\mathcal{A}\mathcal{M}\mathcal{S}$-TeX and the publication in which your paper will appear. Please be sure to include your complete mailing address.

After acceptance. The final version of the electronic file should be sent to the Providence office (this includes any TeX source file, any graphics files, and the DVI or PostScript file) immediately after the paper has been accepted for publication.

Before sending the source file, be sure you have proofread your paper carefully. The files you send must be the EXACT files used to generate the proof copy that was accepted for publication. For all publications, authors are required to send a printed copy of their paper, which exactly matches the copy approved for publication, along with any graphics that will appear in the paper.

Accepted electronically prepared files can be submitted via the web at `www.ams.org/submit-book-journal/`, sent via FTP, or sent on CD-Rom or diskette to the Electronic Prepress Department, American Mathematical Society, 201 Charles Street, Providence, RI 02904-2294 USA. TeX source files, DVI files, and PostScript files can be transferred over the Internet by FTP to the Internet node `ftp.ams.org` (130.44.1.100). When sending a manuscript electronically via CD-Rom or diskette, please be sure to include a message identifying the paper as a Memoir.

Electronically prepared manuscripts can also be sent via email to `pub-submit@ams.org` (Internet). In order to send files via email, they must be encoded properly. (DVI files are binary and PostScript files tend to be very large.)

Electronic graphics. Comprehensive instructions on preparing graphics are available at `www.ams.org/authors/journals.html`. A few of the major requirements are given here.

Submit files for graphics as EPS (Encapsulated PostScript) files. This includes graphics originated via a graphics application as well as scanned photographs or other computer-generated images. If this is not possible, TIFF files are acceptable as long as they can be opened in Adobe Photoshop or Illustrator. No matter what method was used to produce the graphic, it is necessary to provide a paper copy to the AMS.

Authors using graphics packages for the creation of electronic art should also avoid the use of any lines thinner than 0.5 points in width. Many graphics packages allow the user to specify a "hairline" for a very thin line. Hairlines often look acceptable when proofed on a typical laser printer. However, when produced on a high-resolution laser imagesetter, hairlines become nearly invisible and will be lost entirely in the final printing process.

Screens should be set to values between 15% and 85%. Screens which fall outside of this range are too light or too dark to print correctly. Variations of screens within a graphic should be no less than 10%.

Inquiries. Any inquiries concerning a paper that has been accepted for publication should be sent to `memo-query@ams.org` or directly to the Electronic Prepress Department, American Mathematical Society, 201 Charles St., Providence, RI 02904-2294 USA.

Editors

This journal is designed particularly for long research papers, normally at least 80 pages in length, and groups of cognate papers in pure and applied mathematics. Papers intended for publication in the *Memoirs* should be addressed to one of the following editors. The AMS uses Centralized Manuscript Processing for initial submissions to AMS journals. Authors should follow instructions listed on the Initial Submission page found at www.ams.org/memo/memosubmit.html.

Algebra to ALEXANDER KLESHCHEV, Department of Mathematics, University of Oregon, Eugene, OR 97403-1222; email: ams@noether.uoregon.edu

Algebraic geometry to DAN ABRAMOVICH, Department of Mathematics, Brown University, Box 1917, Providence, RI 02912; email: amsedit@math.brown.edu

Algebraic geometry and its applications to MINA TEICHER, Emmy Noether Research Institute for Mathematics, Bar-Ilan University, Ramat-Gan 52900, Israel; email: teicher@macs.biu.ac.il

Algebraic topology to ALEJANDRO ADEM, Department of Mathematics, University of British Columbia, Room 121, 1984 Mathematics Road, Vancouver, British Columbia, Canada V6T 1Z2; email: adem@math.ubc.ca

Combinatorics to JOHN R. STEMBRIDGE, Department of Mathematics, University of Michigan, Ann Arbor, Michigan 48109-1109; email: JRS@umich.edu

Commutative and homological algebra to LUCHEZAR L. AVRAMOV, Department of Mathematics, University of Nebraska, Lincoln, NE 68588-0130; email: avramov@math.unl.edu

Complex analysis and harmonic analysis to ALEXANDER NAGEL, Department of Mathematics, University of Wisconsin, 480 Lincoln Drive, Madison, WI 53706-1313; email: nagel@math.wisc.edu

Differential geometry and global analysis to CHRIS WOODWARD, Department of Mathematics, Rutgers University, 110 Frelinghuysen Road, Piscataway, NJ 08854; email: ctw@math.rutgers.edu

Dynamical systems and ergodic theory and complex analysis to YUNPING JIANG, Department of Mathematics, CUNY Queens College and Graduate Center, 65-30 Kissena Blvd., Flushing, NY 11367; email: Yunping.Jiang@qc.cuny.edu

Functional analysis and operator algebras to DIMITRI SHLYAKHTENKO, Department of Mathematics, University of California, Los Angeles, CA 90095; email: shlyakht@math.ucla.edu

Geometric analysis to WILLIAM P. MINICOZZI II, Department of Mathematics, Johns Hopkins University, 3400 N. Charles St., Baltimore, MD 21218; email: trans@math.jhu.edu

Geometric topology to MARK FEIGHN, Math Department, Rutgers University, Newark, NJ 07102; email: feighn@andromeda.rutgers.edu

Harmonic analysis, representation theory, and Lie theory to ROBERT J. STANTON, Department of Mathematics, The Ohio State University, 231 West 18th Avenue, Columbus, OH 43210-1174; email: stanton@math.ohio-state.edu

Logic to STEFFEN LEMPP, Department of Mathematics, University of Wisconsin, 480 Lincoln Drive, Madison, Wisconsin 53706-1388; email: lempp@math.wisc.edu

Number theory to JONATHAN ROGAWSKI, Department of Mathematics, University of California, Los Angeles, CA 90095; email: jonr@math.ucla.edu

Number theory to SHANKAR SEN, Department of Mathematics, 505 Malott Hall, Cornell University, Ithaca, NY 14853; email: ss70@cornell.edu

Partial differential equations to GUSTAVO PONCE, Department of Mathematics, South Hall, Room 6607, University of California, Santa Barbara, CA 93106; email: ponce@math.ucsb.edu

Partial differential equations and dynamical systems to PETER POLACIK, School of Mathematics, University of Minnesota, Minneapolis, MN 55455; email: polacik@math.umn.edu

Probability and statistics to RICHARD BASS, Department of Mathematics, University of Connecticut, Storrs, CT 06269-3009; email: bass@math.uconn.edu

Real analysis and partial differential equations to DANIEL TATARU, Department of Mathematics, University of California, Berkeley, Berkeley, CA 94720; email: tataru@math.berkeley.edu

All other communications to the editors should be addressed to the Managing Editor, ROBERT GURALNICK, Department of Mathematics, University of Southern California, Los Angeles, CA 90089-1113; email: guralnic@math.usc.edu.

Titles in This Series

946 **Jay Jorgenson and Serge Lang,** Heat Eisenstein series on $\mathrm{SL}_n(C)$, 2009

945 **Tobias H. Jäger,** The creation of strange non-chaotic attractors in non-smooth saddle-node bifurcations, 2009

944 **Yuri Kifer,** Large deviations and adiabatic transitions for dynamical systems and Markov processes in fully coupled averaging, 2009

943 **István Berkes and Michel Weber,** On the convergence of $\sum c_k f(n_k x)$, 2009

942 **Dirk Kussin,** Noncommutative curves of genus zero: Related to finite dimensional algebras, 2009

941 **Gelu Popescu,** Unitary invariants in multivariable operator theory, 2009

940 **Gérard Iooss and Pavel I. Plotnikov,** Small divisor problem in the theory of three-dimensional water gravity waves, 2009

939 **I. D. Suprunenko,** The minimal polynomials of unipotent elements in irreducible representations of the classical groups in odd characteristic, 2009

938 **Antonino Morassi and Edi Rosset,** Uniqueness and stability in determining a rigid inclusion in an elastic body, 2009

937 **Skip Garibaldi,** Cohomological invariants: Exceptional groups and spin groups, 2009

936 **André Martinez and Vania Sordoni,** Twisted pseudodifferential calculus and application to the quantum evolution of molecules, 2009

935 **Mihai Ciucu,** The scaling limit of the correlation of holes on the triangular lattice with periodic boundary conditions, 2009

934 **Arjen Doelman, Björn Sandstede, Arnd Scheel, and Guido Schneider,** The dynamics of modulated wave trains, 2009

933 **Luchezar Stoyanov,** Scattering resonances for several small convex bodies and the Lax-Phillips conjuecture, 2009

932 **Jun Kigami,** Volume doubling measures and heat kernel estimates of self-similar sets, 2009

931 **Robert C. Dalang and Marta Sanz-Solé,** Hölder-Sobolv regularity of the solution to the stochastic wave equation in dimension three, 2009

930 **Volkmar Liebscher,** Random sets and invariants for (type II) continuous tensor product systems of Hilbert spaces, 2009

929 **Richard F. Bass, Xia Chen, and Jay Rosen,** Moderate deviations for the range of planar random walks, 2009

928 **Ulrich Bunke,** Index theory, eta forms, and Deligne cohomology, 2009

927 **N. Chernov and D. Dolgopyat,** Brownian Brownian motion-I, 2009

926 **Riccardo Benedetti and Francesco Bonsante,** Canonical wick rotations in 3-dimensional gravity, 2009

925 **Sergey Zelik and Alexander Mielke,** Multi-pulse evolution and space-time chaos in dissipative systems, 2009

924 **Pierre-Emmanuel Caprace,** "Abstract" homomorphisms of split Kac-Moody groups, 2009

923 **Michael Jöllenbeck and Volkmar Welker,** Minimal resolutions via algebraic discrete Morse theory, 2009

922 **Ph. Barbe and W. P. McCormick,** Asymptotic expansions for infinite weighted convolutions of heavy tail distributions and applications, 2009

921 **Thomas Lehmkuhl,** Compactification of the Drinfeld modular surfaces, 2009

920 **Georgia Benkart, Thomas Gregory, and Alexander Premet,** The recognition theorem for graded Lie algebras in prime characteristic, 2009

919 **Roelof W. Bruggeman and Roberto J. Miatello,** Sum formula for SL_2 over a totally real number field, 2009